Absolute Truth Exposed

Volume 1

Applying Science to Expose the Myths and Brainwashing in the Big Bang Theory, Autoimmune Diseases, IBD, Ketosis, Diet, Nutrition, Dietary Fiber, Carbohydrates, Saturated Fats, Red Meat, Healing, Health, Whole Grains, and the Bible.

Kent R. Rieske, BS/ME

Absolute Truth Exposed
Volume 1

Copyright © 2010 by Kent R. Rieske. All rights reserved.
Absolute Truth Exposed® is a registered trademark by Kent R. Rieske.
Exalt Publishing® is a registered trademark by Kent R. Rieske.
All other trademarks are the property of their respective companies.
Edited by Marti Rieske.

No part of this book may be reproduced or used in any manner by scanning, xerographic, electronics, photographic, or any other means without the written permission of the author. Exceptions are small quotations referencing the source for use in articles or reviews. This book and all of the media, including electronic, photographic, or images of any kind used in the printing of this book shall remain the property of the author, Kent R. Rieske.

Permission granted by the European Space Agency (ESA) to reproduce the image of Vastitas Borealis Crater on Mars.
Permission granted by bigstockphoto.com to reproduce some interior images and all of the cover images except the Bible picture which was taken by the author.
Permission granted by NASA to reproduce images of the International Space Station, Mars, galaxies, descriptions, and all other objects in the universe as public domain.
Permission granted by Tomas Nelson, Inc., to reproduce Bible verses from *New King James Version*, Copyright © 1982.
Permission granted by the Salt Lake City Police Department in the release of pictures related to the kidnapping of Elizabeth Smart as public domain.
Permission granted by the US Federal Bureau of Investigation, Freedom of Information Act for the image of Patty Hearst. Other images are public domain.

Library of Congress Cataloging-in Publication Data
Kent R. Rieske, Absolute Truth Exposed – Volume 1

ISBN-10   0-9828485-1-X   (Paperback)
ISBN-13   9780982848517

1 – HEALTH & FITNESS / Nutrition
2 – SCIENCE / Astrophysics & Space Science
3 – MEDICAL / Diet Therapy
4 – RELIGION / Biblical Criticism & Interpretation / General

First printing July 2010. Exalt Publishing®.
Printed in the United States of America and United Kingdom.

Contents

Introduction		xv
Chapter 1	Absolute Scientific Proof Carbohydrates Are Pathogenic (Disease Causing)	1
Chapter 2	Absolute Scientific Proof Dietary Fiber Is Unhealthy	37
Chapter 3	Top Ten Nutritional Myths, Distortions, and Lies That Will Destroy Your Health	59
Chapter 4	Top Twelve Historical Events That Created Our Current Health and Nutritional Quagmire	131
Chapter 5	Achieving Remission in Asthma, Inflammatory Bowel Diseases, and Other Autoimmune Diseases	169
Chapter 6	Brainwashing, Mind Control, and the Deception of Society	283
Chapter 7	Absolute Scientific Proof the Big Bang and Redshift Light Theories Are Wrong	313
Chapter 8	Bible Interpretation Rightly Divided A Bible Summary	373

Dedication

To my wife, Marti; son, Matthew; and daughter, Michelle.

To the many supporters who have written to me with encouragement and confirmation that the dietary recommendation presented in this book has benefited them greatly, and to those who have expressed sincere thanks, stating that the information has saved their lives.

Acknowledgement

I am very grateful to my wife, Marti, for her excellent editing of this book. Her medical secretarial degree from Concordia College, Bronxville, NY; extensive experience as a medical transcriptionist; previous book editing; and natural command of the English language have given her the unique ability to edit the nutritional, health, medical, and scientific content of this book.

Thank you, Marti, for your wonderful help and continued support.

Medical Disclaimer

All information contained herein is intended for your general knowledge only and is not a substitute for medical advice or treatment for specific medical conditions. You should seek prompt medical care for any specific health issues and consult your physician before starting a new fitness or nutrition regimen. The information contained in this book is presented in summary form only and intended to provide broad consumer understanding and knowledge of diet, supplements, health, and diseases. The information should not be considered complete and should not be used in place of a visit, call, consultation, or advice from your physician or other health-care provider. I do not recommend the self-management of health problems. Do not self-diagnose your health condition based on the information provided in this book. Seek a diagnosis from your doctor. Information obtained by reading this book is not exhaustive and does not cover all diseases, ailments, physical conditions, or their treatment. Should you have any health-care related questions, please call or see your physician or other health-care provider promptly. You should never disregard medical advice or delay in seeking it because of something you have read here. I strongly suggest you select a physician who is knowledgeable and supportive of the low-carbohydrate lifestyle.

The diet program presented in this book is not to be interpreted as a cure for any disease. The diet information presented in this book is based on scientific analysis that show some foods cause negative reactions in some people and other foods do not.

I must emphasize that the criticism of some doctors, nutritionists, teachers, professors, scientists, literary agents, publishing companies, media companies, politicians, and others should not be viewed as a blanket criticism of all. Many truly awesome people in these industries are providing a valuable service.

Autobiography

**Kent and Marti Rieske
Hallet Peak and Emerald Lake
Rocky Mountain National Park, Colorado, USA**

I was born at home on a small farm in Provo, Utah, in 1939. We grew fruit, vegetables, and grains in abundance. The modern food shopper would be astonished at the amount of food we threw away or let rot. Most of the farm products were naturally organic, but we sprayed the apples and beans with pesticides several times each season. Worm-free organic apples are impossible to grow, and those sold in the supermarkets are a fraud. My mother prepared meals from this abundance, and my father loved the cornbread made from home-grown, home-ground corn. This diet was the mirror image of the *USDA Food Guide Pyramid*. My mother died from cancer at age 50 and my father from heart disease at age 56 as a result of this high-carbohydrate diet.

I graduated from college in 1962 with a Bachelor of Science degree in Mechanical Engineering and became a registered Professional Engineer in several states. My career included positions as Chief Mechanical Engineer and Director of Engineering. In 1985, I started a home-based consulting engineering firm, developed and marketed engineering software worldwide, and created a non-denominational Christian website based on dispensational theology. The website also includes a scientific approach to health and nutrition.

In the early 1990s, my wife and I started eating lots of fruits and vegetables combined with homemade seven-grain muesli and home-ground whole-grain breads—a mirror image of the official food guide pyramid. This was the perfect "balanced diet," but it nearly killed both of us. I developed high blood pressure at age 44, which I later found was caused by arterial plaque. By the way, **Absolute Truth Exposed – Volume 2** will tell you how to reverse heart disease. This first volume simply does not have the space.

After reading *Dr. Atkins' New Diet Revolution,* I began my research into the relationship between diet and health because the balanced diet made me very fat and gave me heart disease. I also received excellent information from books by Dr. Michael R. Eades and his wife, Dr. Mary Dan Eades. *Protein Power* and all of their other books are excellent. Switching to this new low-carbohydrate diet made an awesome improvement in my health, and I lost 40 pounds.

My extensive 10-year scientific study of nutrition, health, medicine, and human physiology gave me a depth of knowledge rarely achieved even by those educated as professional nutritionists or earning more advanced Ph.D. degrees. Physicians are usually not good sources for nutritional advice because they receive very little nutritional training in medical school.

My research revealed the shocking truth that officially-promoted nutrition, as taught in universities and sponsored by government agencies, is nothing more than brainwashing with myths, distortions, and lies. The *USDA Food Guide Pyramid* has no connection with scientific facts.

Over a period of three years, I monitored an Internet message board that specialized in digestive ailments. Although the diet promoted on the message board provided some help for the sufferers, it was severely deficient. Three or four of the 1000 members recognized that numerous foods from the acceptable list were actually offensive, but the organizers of the message board quickly dismissed their reports. These three or four people restricted their diets enough to enjoy a reasonable remission from Crohn's disease and ulcerative colitis. The diet promoted by the board's founder was wrong, and the remaining 99.6% continued to suffer.

The proper diet required to prevent harsh autoimmune reactions is generally not recognized by gastroenterologists (physicians who specialize in digestive disorders). They continue to recommend dietary fiber that my research revealed to be the main culprit. The high-fiber diet makes patients' health worse—not better. Hence, many gastroenterologists have given up recommending dietary changes to their patients and rely solely on drugs, which are intended to destroy the immune system rather than stop the offenders—fiber and carbohydrates.

My research has also revealed that our society has been brainwashed with false information on nearly every topic of concern to our everyday lives. Discovering the truth is the connecting link between the title of this book and nutrition, health, medicine, science, and religion. Yes, a strong connection exists between brainwashing, religion, and diet in our society.

I placed my research information on a website that received 3,000 to 4,000 thousand visits per day, and I provided my email address to give the visitors a chance to share their personal experiences. This is directly opposite to the attitude of most physicians, who simply don't want any feedback from the patient. The feedback from the website was a perfect confirmation of the research information. Dietary fiber is a sinister invader of the colon, not the panacea of health promoted in other books, websites, universities, and government health departments. Fiber is a bad dude.

Carbohydrates were found to be as bad an actor as fiber. There are no good carbohydrates as popularly claimed. Some carbohydrates are bad and the rest are dreadful. I will provide scientific proof to support my statements.

I hope you have the courage to read and study this book in detail. The material has been condensed and summarized to give you a jump-start into good health. Many diseases can be reversed or placed into remission by adopting my recommendations for diet, supplements, and standard medical support. I strongly recommend that you seek medical care from a physician who truly understands the connection between diet and health. Simply reading this book won't bring health improvements. You must be proactive and make significant changes to your lifestyle.

Credits

William Banting (1797 - 1878) was a formerly obese undertaker in London, England, who wrote *Letter of Corpulence*, a 22-page booklet, in 1859, in which he explained how a fairly low-carbohydrate diet greatly restored his failing health in a very short time. His success was based on his persistence in trying every diet suggested to him by several doctors until he stumbled upon the absolute truth regarding the connection between diet and health. To his dismay, the information was instantly rejected by the majority of physicians and citizens alike.

Vilhjalmur Stefansson, BA, ethnologist (1879 – 1962), and **Rudolph Martin (Karsten) Anderson**, PhD, zoologist (1876 – 1961), were Arctic explorers who ventured to Northern Canada beyond the Arctic Circle in 1906 – 1912 to bring back the secrets of the Eskimos' all-meat diet. Mr. Stefansson was shocked upon his return at the rejection his report received from the public and doctors alike.

Weston A. Price, DDS, dentist and nutritionist (1870 – 1948), traveled around the world with his wife, Florence, in the 1920s and 1930s, visiting many primitive villages in order to assess the connection between diet and their excellent health. He authored the popular book, *Nutrition and Physical Degeneration.*

Robert C. Atkins, MD, cardiologist and nutritionist (1930 – 2003), authored the best-selling low-carbohydrate book, *Dr. Atkins' Diet Revolution,* and several others.

Michael R. Eades, MD, and **Mary Dan Eades**, MD, doctors and nutritionists who authored the very popular low-carbohydrate book, *Protein Power,* and several others.

Absolute Scientific Proof Will Be Presented in This Book to Prove the Big Bang Theory Is False

Absolute Truth Exposed

Volume 1

Introduction

You are about to enter the pristine world of **Absolute Truth Exposed**. Truth is being suppressed in all areas of our everyday lives, including science, nutrition, health, medicine, religion, and others. The scientific connection between diet and health is major area of mass deception. Society is being brainwashed by food manufacturing companies, health practitioners, universities, media groups, religious groups, political parties, governments, and other biased organizations.

We have all been brainwashed by the obsessive compulsive controlling élite. Are your thoughts correct and true or has the élite gotten to you? You will never know. We think we do. A person will say, "Everyone else is wrong—not me." That person is the one most likely to be brainwashed because he doesn't understand the threat. He just soaks up false information like a sponge. That is the nature of brainwashing.

Introduction

I don't expect people to agree with much of the information presented in this book, but I can understand. That's the nature of absolute truth. It is human nature to believe everything but the truth. I will give examples of people believing lies for centuries.

The human species has a very destructive mental, spiritual, and psychological weakness in its susceptibility to deception. This weakness is obvious in our everyday world where people strongly disagree on every subject imaginable. There is no consensus on any subject, and we all can't be right. Perhaps none of us is right, but this book will attempt to sort out the prejudices, agendas, political correctness, monetary trickery, and religious and atheistic nonsense to arrive at the absolute truth by strict scientific evidence. The nutritional claims made in this book have been tested by people who are seeking relief from ailments that modern medicine could not provide.

This book has several topics but one main theme, *Absolute Truth Exposed*.

The one major theme of this book is to expose truth wherever it can be found without regard for the fact that each individual in our society has widely differing opinions. Each chapter is based on a separate, widely diverse subject. The topics are summarized and condensed without the endless fluff and bulk found in most books. Each chapter could easily have been made into a separate book as is commonly done. This unusual format gives you condensed information equivalent to eight books. Each differing chapter feeds into the main theme of the title. Absolute truth does exist, but it is hidden beneath the mass of false information thrust upon society. Those who find the truth will receive rewards that are almost miraculous.

Introduction

The far-reaching topics include nutrition, health, medicine, science, psychology, and religion. Each chapter identifies a topic that challenges the commonly-held beliefs found in our society. The absolute truth is presented to correct these false beliefs. Because misinformation is rampant, you will most likely disagree with the conclusions presented in some of the chapters. Books, websites, organizations, and schools have filled people's mind with so much misinformation that readers will find it almost impossible to believe the claims made in this book.

We are a society of severely brainwashed people. We shouldn't be surprised at this statement. Much of what people believed in past civilizations was not true. They believed the Earth was the center of the universe. Oops! People who believe the Big Bang Theory still do. (That was not a good example.) In past centuries, people believed the Earth was flat. More recently, some astronomers and others believed there were aliens living on Mars. The Eskimos believed their Shaman (spiritual leader) could travel to the Moon and talk to the people.

The argument over right and wrong has collapsed to the point where some people claim nothing is right and nothing is wrong. People are told to establish their own reality. Whatever they believe will be right for them. This is nonsense. Absolute truth exists, and we should diligently seek to find it.

This book is not intended for a niche of like-minded supportive readers as is the case with many books. Readers who vehemently disagree will hopefully give the topics more thought. The reader who is always searching for the hidden truth will find this book compelling, while the reader with a blocked mind will find it frustrating since **Absolute Truth Exposed** clashes with chiseled-in-stone false beliefs.

Introduction

We have become very skeptical as a society, and you most likely will be skeptical about the conclusions presented here. We are always on guard and fearful of being brainwashed again. We are drawn into a philosophy by people with high credentials who promise great results. We desperately want the desired result, so we convince ourselves the program is working. Slowly the truth sinks in. We must admit the facts. We have been duped again. The program was destructive—it did not provide the great results we were promised.

When I asked a bank teller if she believed in absolute truth, she answered, "No." My next question was, "Are you absolutely sure?" Her brain literally locked up, and her eyes faded into a blank stare as she pondered the question that was never answered. She would have to admit that absolute truth may possibly exist if she answered, "Yes," and she would be admitting that absolute truth was a possibility if she answered, "No." She could not say that she was absolutely sure. Her brain could not arrive at an acceptable answer that would be in agreement with her new core belief. She said nothing because she did not think absolute truth was a possibility, but she wasn't sure.

Brainwashing—defined as mind control, coercive persuasion, and thought reform—leads us to believe something is true when it is actually false. People are very easily brainwashed. In fact, brainwashing is so prevalent in our society that advertisers use the technique on a regular basis to sell their products. We can't prevent ourselves from being brainwashed no matter how hard we try. Many people make little attempt to protect their brains from being controlled by others; they simply open their brains to the world around them. But the majority of people appear to have developed shields, barriers, blockades, and guards to protect thoughts that they regard as absolute truth. They protect

Introduction

these extremely precious thoughts even though they may not be truth at all. They migrate to like-thinking people, groups, and institutions as safe havens for their brains. The mental anguish about life is easier to endure when they do not have to be on constant guard against objectionable ideas and destructive thoughts sinking deep into their mental processes.

The awesome evidence in this book proves the Big Bang Theory is wrong.

This book contains one of the most important scientific discoveries made in your lifetime. I call this new discovery the *Rieske Spacetime Drag Coefficient for the Speed of Light*. This proclamation was first made worldwide on an Internet website on November 8, 2005. Dr. Albert Einstein's *General Theory of Relativity* was used to validate this new discovery. I make the claim that the redshift in the spectrum of light coming from distant galaxies is caused by a decrease in the speed of light, not by expansion of the distant galaxies accelerating away from Earth. The *Rieske Spacetime Drag Coefficient for the Speed of Light* proves beyond any possible doubt that Dr. Edwin Hubble's *Redshift Light Theory* is wrong and in doing so, destroys the *Big Bang Theory* as well. This new Law of Science proves that the speed of light slows as it travels through the universe due to the resistance imposed on mass by spacetime. The decay in the velocity of mass traveling through spacetime has already been confirmed by observing the decay in the orbital time of binary pulsar PSR 1913+16. The distant galaxies are not accelerating away from Earth as astrophysicists and cosmologists have claimed. The speed of light is not a constant as stated in science books and taught by university professors. The universe is not in a state of accelerating expansion. The Big Bang Theory is

Introduction

destroyed by the *Rieske Spacetime Drag Coefficient for the Speed of Light*.

Do you have an autoimmune disease, such as asthma, ulcerative colitis, Crohn's disease, rheumatoid arthritis, psoriasis, or lupus? Keep reading. The diet program presented in this book can place your disease into remission almost like a miraculous healing. Your doctor will strongly object to this diet program because we are all being brainwashed.

> **You are about to enter the world of Perfect Diet —— Perfect Nutrition.**

Do you think a diet with lots of fiber is healthy? Keep reading. The human body cannot digest dietary fiber, but pathogenic bacteria and yeast in your colon absolutely love it. They produce toxic byproducts that will make you sick. The high-fiber diet has been promoted because we have all been brainwashed.

History is awash with enormous tragedies that result from individual and mass brainwashing. The first obvious example is Germany's submission to brainwashing by Adolph Hitler. He was elected Chancellor of Germany by a popular democratic vote, thus absolutely proving that democracy is not a sound method for selecting leaders. The people were seduced into the worship of Hitler as observed by their salute, "Heil Hitler." Brainwashing increased to the extreme. Citizens who had previously been of sound moral conviction began rounding up members of minority groups for induction into forced slave labor or taking them to the slaughter chambers. Hitler brainwashed the citizens into believing that Germany could conquer the world and become master of the entire planet. This was obviously nonsense, but the people believed it because they were all brainwashed. The

Introduction

absolute military truth was evident even then. Germany's defeat would be inevitable. She didn't have enough people, resources, or isolation to expand far beyond her immediate neighbors and soon hit the wall of reality as she tried to move beyond the motherland in her attempt to defeat Russia. The truth was coming home to Germany as the survivors dragged their half-frozen bodies back to the homeland.

Failure of the human mind to easily discern truth from error is very costly to both the individual and the larger society. We fall prey to scam artists on a regular basis even though we all appear to make some attempt to guard against it. We groan in disgust with ourselves as we realize that a slick promoter with a bunch of lies has tricked us again. We constantly seek absolute truth, but we rarely identify it. Those who claim absolute truth does not exist are actually holding to that belief as if it were the absolute truth. They have given up. Unable to discern truth from error, they simply say truth does not exist.

Erase your brain now and keep reading if you want *Absolute Truth Exposed*.

Many people consider a favorite political, religious, or educational establishment to be their guide and seek out those members or teachers who protect them from outside brainwashing. This approach does not work well because they are simply opening their minds to inside brainwashing. It would be better if they were suspicious of everyone. They drift deeper and deeper into the same mindset as the group until they reach the point of no return. These groups often withhold their core beliefs from newcomers because they plan to slowly indoctrinate them. This is the primary brainwashing strategy.

Introduction

I encourage you to be suspicious of the claims made in this book. Test them yourself. Test the claims in the chapter on autoimmune diseases. Record the frequency of your inhaler use if you have asthma. The diet presented will place your asthma or irritable bowel syndrome into remission. Your inhaler use will drop sharply and most likely be discontinued completely. Don't reject the claims summarily without testing them yourself. Record your inhaler use while on the diet. Compare the results with the effectiveness of the treatment prescribed by your doctor. You can discover the truth yourself.

We become locked into a belief system that usually lasts for a lifetime because the human mind appears to lack the ability to reverse course and will generally not recognize that it has been brainwashed. This dreadful state leaves the individual with little hope of reversing the condition. Everyone believes differently, but no one believes he has been brainwashed. Families have been known to use extreme and often unlawful rescue tactics in their attempts to recover other family members from the grasps of seductive brainwashing organizations or individuals, but these attempts often fail.

We commonly obtain truth by depending on others who have a high level of education or experience. Book authors and organization leaders are constantly being asked about their credentials; this gives the false impression that a high level of education equates to a high level of truth. The opposite is most likely the case. It is highly probable that the person with credentials has a high level of brainwashing.

The Wright brothers were the first people to design, build, and fly an airplane. They are examples of great achievements by those without credentials or education. They owned a bicycle shop where they made and sold bicycles. They did not have

Introduction

advanced degrees of higher education. They had no credentials, yet their accomplishments were extraordinaire. Their one divine principle that drove them to succeed where those with advanced credentials failed was their effortless search for the truth regarding airplane design and construction. The airfoil shape for the wing was adopted by mimicking the shape of a bird's wing. They invented a wind tunnel even though they had no education in fluid dynamics, a subject still in its infancy. The wind tunnel was not used to design the wing as one might assume. It was used to design a propeller that would provide the highest thrust-to-power ratio. Modern aeronautical engineers call this the *lift-to-drag ratio* that gives a measure of efficiency. The Wright brothers could not find a powerful lightweight engine, so they designed and built their own four-cylinder engine.

Societies in English-speaking countries have been changed by the advertising industry. Advertisements have trained people to expect and desire a mental "sales job" before they make a decision. People want to be persuaded instead of seeking the truth for themselves; they are begging to be brainwashed. The advertising industry uses innuendoes, subtle suggestions, and appeals to the emotions rather than presenting information in matter-of-fact, direct statements. Examples can be seen in automobile advertising in which emotional appeals are used to attract buyers. Claims are made that the car brings out the tiger in you, offers high adventure, helps to save the planet, or provides the ultimate in safety features. Advertisements for a Dodge heavy duty truck never tell you that the connecting rod bearings in the Cummins straight six-cylinder engine are larger than on the competing V-8 engines. The larger bearing will last longer before the engine needs repair. That is a solid engineering fact that I want to hear, but I can't see these facts in a television advertisement. I must search out the truth myself.

Introduction

Our society often resents factual statements. People who make matter-of-fact statements are viewed as having chips on their shoulders or being know-it-alls. This book does not use the subtle suggestion method of gentle persuasion. It is direct and to the point. Statements are made without any attempt to do an emotional sales job on the reader. References are rare in this book because original thought has no reference. Our society has grown to believe that original thought does not exist. Books are plastered from beginning to end with references to authors who reference other authors until the circle of references forms an endless link from author to author. This approach gives the false impression that the topic is valid. Readers who disagree with the material presented are quick to demand a reference but never bother to check the ones provided. Much of the information released to the public by professional organizations is absolutely false and so are the references. Having wide support and extensive references do not make a false topic true.

Brainwashing and the denial of absolute truth can persist for centuries. The death of US President James A. Garfield in 1881 resulted from a pistol shot. He would probably have recovered had his team of Army doctors not treated him. Many people have recovered and healed very well with bullets remaining lodged in their bodies. In the end, the doctors managed to take a three-inch wound and turn it into a twenty-inch canal that was heavily infected and oozed more and more pus with each passing day because they probed the bullet wound with unwashed hands and dirty instruments. Medicine had rejected the theory of bacterial infection even though bacteria had been discovered more than 200 years earlier. In 1676 – 1677, Antonie Philips van Leeuwenhoek, a Dutchman, was the first to record microscopic observations of bacteria and spermatozoa. Bacteria could be seen in infected flesh wounds, whereas fresh uninfected wounds

Introduction

could be seen to be bacteria free. It seems ridiculous that highly-credentialed physicians could deny this simple truth for 200 years, but that is the nature of brainwashing. It's hard to imagine but true. For ten generations, older physicians and medical schools brainwashed new students into rejecting the truth that invisible bacteria cause infections. Scientists, professors, and physicians are not immune to delusional thinking and false beliefs. Scientists with high credentials are teaching in the world's top science schools, and they will continue to brainwash new students for many generations to come regarding Dr. Hubble's Redshift Light Theory and the Big Bang Theory even though they are totally false. You may come to realize the Big Bang Theory is false after reading this book, but I doubt it. Remember, we are all brainwashed. Brainwashing is almost irreversible. You are doomed to live your entire life believing a galactic lie.

Have you ever been deceived by a person with outstanding credentials?

Credentials do not provide a person with absolute truth. People of equally high credentials often arrive at opposite views and opinions. It's a mystery why people insist that information come from a source with high credentials when many of those with equally high credentials disagree. Credentials generally mean the person has spent many years engaged in advanced education. This doesn't mean they have been learning the truth. More than likely, they have an advanced degree in submission to brainwashing. There is a big difference between knowledge and wisdom.

Health authorities tell us to eat a balanced diet with 50% to 60% carbohydrates. They don't tell us that our health would be better

Introduction

if our carbohydrate intake were zero. This book presents the absolute scientific proof that carbohydrates are unhealthy. You have been brainwashed if you believe the *USDA Food Guide Pyramid*. Carbohydrates are not an essential macronutrient in the human diet. Why are we being told to eat 60% of something that we don't need? Diabetes is a raging epidemic because of this incorrect advice. Eskimos lived on the frozen Arctic Ocean for nine months of the year eating only seal meat and blubber (fat). Their diet was at least 80% fat and their health was excellent. The health authorities are not telling us the truth. You will read the Absolute Truth in this book with the science to prove it.

Health authorities also tell us to eat a high-fiber diet. The Eskimo diet had no fiber and no carbohydrates. Meat does not contain fiber. Fat does not contain fiber. Something is amiss. This book will tell you the truth. Dietary fiber is very unhealthy, and this book will explain in detail the reasons why.

The *USDA Food Guide Pyramid* is scientifically false.

Ten major myths have caused our current health and nutritional quagmire. I expose all of them and present the Absolute Truth. Fruits, vegetables, and whole grains are not required in a healthy diet. For centuries, the ancient Egyptians living on the Nile River Delta grew abundant crops with ease. Fruits, vegetables, and whole grains were the mainstay of their diet—exactly the same as recommended in the *USDA Food Guide Pyramid*. The results were disastrous. Egypt has more than 100,000 mummies to study. They had heart disease, cancer, diabetes, degenerating spines and joints, severe dental decay, rheumatoid arthritis, and the entire list of other autoimmune diseases. Today we are duplicating the fate of the mummies.

Introduction

Before each mission to explore Mars, NASA makes a proclamation that they hope to find water that will contain a life form. We all wait anxiously for the result from the Mars module as it digs into the surface and performs the tests.

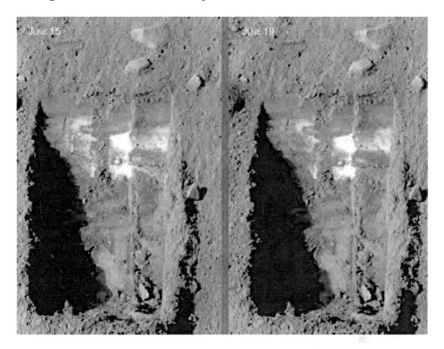

Soil Marks on Mars Made by the Lander Scoop

The NASA Phoenix Mars Lander touched down in 2008. On June 15, a scoop device retrieved a soil specimen for the test module to analyze. The mass media frantically made predictions that finding water near the surface would lead to the discovery of life on another planet.

Wake up! This is brainwashing. NASA's test results were no surprise to me. Mars is absolutely sterile.

Several years earlier, the European Space Agency (ESA) Mars Express satellite took a picture of a frozen water lake in a crater on Mars. Yes! A lake! NASA claims Mars has very strong seasonal

Introduction

winds, but the picture shows the surface of the lake to be free from the slightest dusting of silt. NASA is the epitome of credentials. They have rooms full of people with the highest credentials. How many Ph.D.s does it take to see a lake? Why do they keep searching for water on Mars? Why don't they land on the lake if they want to find water?

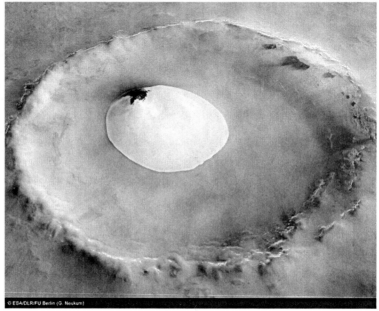

Vastitas Borealis Crater on Mars

The ESA Mars Express picture was taken on February 2, 2005. Three years later, NASA was still searching for water near the surface and raising hopes that life on Mars would soon be discovered.

Do you have the strange feeling NASA is playing with your brain and wasting your money? We can see that no vegetation has grown along the edge of the lake. Snow geese are not circling the crater looking for a good landing spot on the ice. The frozen water is not green from the remnant of algae that filled the lake

Introduction

during a previously warmer climate. The ice is flat, proving it was once liquid water and has since frozen. It is not chunks or blocks of ice. The surface is parallel to the surface of Mars, indicating the stability of the planet. Let us accept the absolute truth. Mars does not have any life forms. It is absolutely sterile. Don't be brainwashed by NASA. The second volume of **Absolute Truth Exposed** will discuss this hoax in greater detail. There is more to the story.

In 1929, Dr. Edwin Hubble and his assistant, Mr. Milton Humason, observed that light from the most distant galaxies had a shift to the red side of the spectrum. They promptly concluded the redshift was caused by the galaxies moving away from Earth at a high rate of speed and theorized that the length of the light waves was longer due to the high velocity and acceleration of the galaxies. They reasoned that the redshift in light phenomena was similar to the Doppler effect in sound waves coming from a moving source. The conclusion that the redshift of light was caused by an expanding universe became known as the *Hubble Redshift Light Theory*. Dr. Hubble's conclusion was received immediately and unanimously accepted by astronomers, astrophysicists, and cosmologists without any further scientific proof. The Redshift Light Theory quickly led to the Big Bang Theory with virtually no other scientific proof. The quick assumption was made that all of the distant galaxies must have come from a central point of origin if they are now all moving away from Earth at a high rate of speed. All of the velocity vectors point directly toward us.

We should always be suspicious when someone tells us that we are the center of the universe.

Introduction

We must ask ourselves a simple question. Did the Big Bang throw all of the other galaxies out into the universe while leaving our Milky Way galaxy and Earth still at the center? You are brainwashed if you said, "Yes!" Image a massive hydrogen atomic bomb exploding over your city. Not only that, but the bomb explodes directly over your house. After the dust settles, you realize all of the houses in the city were blown entirely away except for your house, which sits in the center of the big bang totally unscathed. The Milky Way Galaxy did not get left behind as Big Bang residue. I know many believe the Big Bang Theory is true, but I will prove the Hubble Redshift Light and Big Bang Theories are total nonsense.

Hubble's conclusion has never been seriously challenged until now, and this challenge is not based on a religious alternative as many would expect. The Biblical description of creation could fit into a Big Bang scenario. Many religious people consider that Genesis 1:1, the first verse of the Bible, describes the Big Bang.

> Genesis 1:1 "In the beginning God created the Heavens and the Earth."

Many other verses in the Bible, such as Jeremiah 10:12, clearly describe an expanding universe.

> Jeremiah 10:12 "He has made the earth by His power, He has established the world by His wisdom, And has stretched out the heavens at His discretion."

Chapter 7 will explain the correct scientific reason why light from distant galaxies has a redshift, and it is not the Doppler effect as assumed by Dr. Hubble. The correct reason that light from distant galaxies has a redshift in the spectrum is in complete agreement with Dr. Albert Einstein's *General Theory of Relativity*. In fact, Einstein's theory gives us the correct answer.

Introduction

The scientific truth presented here destroys the Big Bang Theory as well. Society will still cling to the Big Bang Theory because they feel it replaces God as Creator, and they must have a theory for the origin of the universe without a creator. This book will not result in changes to college curricula. It may take 200 years before university professors accept the fact that the Big Bang Theory is a dead corpse. It also required two centuries for physicians to accept the fact that bacteria caused infections. Remember, brainwashing is almost irreversible. Your professor will be teaching the Big Bang Theory until his retirement. His great grandson will still be teaching the same lies about the Big Bang Theory when he retires.

The Big Bang Theory is a dead corpse, but your university professor will still put makeup on it.

Each chapter in this book presents a topic we encounter in our everyday lives. The false information in each topic is challenged, and the truth is presented. The truth is not simply my opinion but is based on hard scientific facts that our modern society ignores in the same manner that physicians denied and ignored the relationship of bacteria to flesh infections for 200 years and NASA refused to accept the beautiful frozen lake on Mars. Each of the topics can have a profoundly beneficial effect on your life if you are one of the precious few who succeed in breaking the brain fog barrier called *brainwashing*.

**The Big Bang Theory was never true.
We have been brainwashed again.**

Seeking truth is vitally important because false beliefs cause hardship, poor health, and a shorter life. Obtaining and embracing absolute truth produce results that appear to be

Introduction

almost miraculous. Each chapter in this book is good example of the brainwashing of our society, but one of the best is Chapter 5 titled, "Achieving Remission in Asthma, Inflammatory Bowel Diseases, and Other Autoimmune Diseases." Inflammatory bowel diseases (IBD), asthma, rheumatoid arthritis, and about thirty other diseases are autoimmune diseases that all respond to the diet in a similar manner. The diet program presented is effective in reducing the immune system's attack on the body by each of the diseases.

Inflammatory bowel diseases such as ulcerative colitis and Crohn's disease are dreadful conditions. In severe cases of ulcerative colitis, a chronic case of constipation turns into raging diarrhea and bleeding from the ulcerations in the colon. Living with constipation is not seriously disruptive to one's lifestyle. Although a person can work, shop, and go to parties and sports events while constipated, this is not the case with raging, bloody diarrhea that practically forces the sufferer to become homebound. Modern professional medical societies claim the cause for IBD is unknown and there are no cures. This book will present the cause and provide easy lifestyle changes that will place the dreaded illness into remission. Unfortunately, IBD is an autoimmune disease in which the immune cells incorrectly attack the tissues of the body. The immune system has a long memory, so a lasting cure is not possible without somehow wiping out the immune system's memory. That may be possible someday, but for now, the diet program contained in this book is the best way to achieve remission.

The correct diet for people with an autoimmune disease is almost like a miracle cure. The digestion returns to normal, and the symptoms disappear. The diet requires 100% compliance in order to prevent harsh flares from occurring at any time. Flares

Introduction

are triggered by eating the wrong foods, contracting another disease like the flu, or having an emotional crisis. The flare will subside in a few days or weeks as the sufferer learns how to manage the disease. The temporary use of a prescription drug often helps to calm the flare, but these drugs will never overpower the severely negative response from eating one little bite of a forbidden food. The immune system has a very long memory, so these autoimmune diseases are never truly cured. Achieving a solid degree of remission is the goal.

> **Normal digestion can be restored to IBD sufferers by adopting the proper diet as presented here.**

Gastroenterologists are doctors who specialize in digestive disorders. Unfortunately, the standard diet they recommend is absolutely backward, and most of the foods will cause a person who is in remission on my diet program to immediately explode into a flare. Other diet programs, such as the vegetarian, gluten-free, or Mediterranean, will cause an immediate flare as well in people with a serious autoimmune disease. Those diets and others are recommended in many books, doctors' websites, and by prestigious organizations, but they simply do not benefit the sufferer. This has led most gastroenterologists to falsely claim that special diets are not helpful for the IBD sufferer. One sufferer, who was well into remission from ulcerative colitis by following the guidelines in the IBD chapter, visited his brother in Italy. Out of respect for his brother, he ate lasagna for dinner. The result was disastrous. He immediately went into a raging flare. The diet in this book had restored his digestive health, and he had gained much-needed weight and muscle. Violating the rules resulted in a dreaded outcome as the immune system attacked his digestive tract again.

Introduction

One person had a second colonoscopy (visual examination of the colon) one year after starting the special IBD diet recommended in this book. Her digestion was perfectly normal during most of that period. The gastroenterologist stated during a review of the second colonoscopy that he would not have known she had Crohn's disease if he had not observed it himself during the first colonoscopy. Absolute truth does indeed perform miracles when it can be found.

I was experiencing pain and tingling in my neck, and my doctor ordered an x-ray. His diagnosis was disturbing. Cervical discs C4, C5, and C6 had thinned which caused the vertebrae to move closer together. The vertebrae had begun to squeeze the branch nerves that extend from the central spinal cord into the arms and hands. He recommended that I see a surgeon who had performed an operation on his neck for a similar condition, but I hesitated to make the appointment. Neck surgery near the spinal cord and the branch nerves is a scary thought. Instead, I researched the subject to find the possible cause and developed a list of actions that had the potential to reverse the thinning discs. Observable improvements could be felt in my neck within a few months after implementing the list of actions, and my neck was perfectly normal a year later—without surgery. Doctors cannot do that. None of these actions is taught in medical school, but you can implement all of them by following the protocol outlined in detail in a future volume of **Absolute Truth Exposed**.

You may have been brainwashed into thinking that a 70% fat diet will kill you tomorrow, or the next day for sure. Certainly no one has told you it was a part of an amazing diet program to reverse heart disease. The American Heart Association (AHA) talks about saturated fat as if it were poison. I switched from the AHA high-carbohydrate diet to a diet containing 70% fat, 30% of

Introduction

which was saturated. The protein in the diet was 24% and the carbohydrates a meager 6%. Fiber was essentially zero. I was fairly active but only exercised strenuously during a quick 20-minute bicycle ride around the neighborhood. My cardiologist was very concerned about my total cholesterol at 241 and LDL cholesterol at 168, although my HDL (good) cholesterol was a respectable 57. He prescribed 20 mg of Zorcor (simvastatin) to lower my cholesterol, but I had heard it can lead to muscle weakness and abnormal liver function tests. After a period of investigation and self-testing, I developed my own effective protocol that provides awesome results—2000 mg of over-the-counter time-release nicotinic acid with a 10 mg boost of simvastatin, taken daily at 8:00 PM. My cardiologist was thrilled at my new cholesterol readings after one year on the program. Total cholesterol plunged from 241 to 157, and the LDL cholesterol plunged from 168 to 75. Wow! The good HDL cholesterol soared from a good reading of 57 to a much better reading of 68. All of the other blood tests were in the normal range. The protocol had no negative side effects and left me feeling strong and energetic. Berkeley HeartLab, Inc. performed cholesterol measurements consisting of the various fractions of the LDL and HDL cholesterol, and the results indicated that my LDL had a large percentage of the safe, large molecule fractions, not the unhealthy small variety often called *BB* type.

This isn't the end of the reversing heart disease story. The 70% fat diet and the self-developed cholesterol-lowering protocol have resulted in a measured reduction of two small plaque inclusions in my heart arteries. My cardiologist measured the two areas of plaque during the first angiogram and diagnosed them as 50% and 40% restrictions in two arteries. These are not life threatening unless a rupture occurs and forms a blood clot. This clot can be prevented by taking 81 mg of aspirin with

Introduction

breakfast and 75 mg of Plavix® with dinner. When he performed another angiogram one year after I started the cholesterol lowering protocol, my cardiologist measured the plaque as only 25% and 20% restrictions—a 50% reduction in artery plaque. Truth has won again. Heart disease can be reversed. I am hopeful that volume two of **Absolute Truth Exposed** will be able to present all of the scientific facts and test results to prove it. Please be patient.

Baby brains are not made from fruit juice and lettuce.

Doctors tell pregnant women to avoid gaining excessive weight during pregnancy. They are told to eat a balanced diet with lots of fruits and vegetables and take a daily prenatal vitamin tablet. Pregnant women are generally very compliant in following this advice, yet infant autism in all English-speaking countries is soaring. The epidemic has reached such a ridiculously high rate that one child in every one hundred is autistic. Why is this happening? The medical authorities admit they have no idea, but the answer is easy to discern when one goes back to elementary Human Physiology 101. Baby brains are not made from fruit juice and lettuce. The developing brain of the baby in the womb is made from cholesterol, fat, and protein. These are the very foods Mommy has been avoiding so she won't gain weight during the pregnancy. The baby didn't gain much weight either, and its low birth weight of only five pounds and small head required that it be placed on immediate life support. Brainwashing has given Mommy and society another autistic child. This first volume of **Absolute Truth Exposed** simply did not have space to expose the dreadful diet information given to pregnant women, but it will be presented in a future volume of **Absolute Truth Exposed**.

Introduction

This book is not time sensitive. It can be kept for future study to glean information needed to treat a new illness or solve a new problem encountered in everyday life. You will never know when a life-threatening crisis will loom before you. The answer you need could be in this book.

> **Many books have 150 pages of recipes, but this book has only one. See page 281.**

I have attempted to summarize and condense as much information as possible in this book. The information is not loaded with endless verbiage in an attempt to increase the page count. I have read some nutrition and diet books that are more than 50% common, uninteresting recipes. I am not against recipes, but this book simply did not have space for them.

I have presented very detailed diet information that leaves little doubt as to which foods are acceptable to eat and which are not. Strict adherence to the diet is vitally important for anyone who has an autoimmune disease, such as asthma, ulcerative colitis, Crohn's disease, psoriasis, lupus, multiple sclerosis, or one of the many others because eating one bite of food from the **Forbidden Foods** list will immediately result in a flare (a severe autoimmune system attack). People who have strictly applied this diet information have testified that they have achieved remission in the symptoms of their autoimmune diseases.

The first chapter could very well be a shock to your dietary understanding. Carbohydrates are touted as the primary energy macronutrient that you must have daily. I hope the scientific data presented in this book will convince you that carbohydrates are not the innocent energy source most professional nutritionists

Introduction

have made them out to be. I will give you solid scientific reasons why all carbohydrates cause disease.

Chapter 2 presents the absolute truth about dietary fiber. No, it is not the panacea for optimum digestive health. In fact, I will prove dietary fiber to be one of the unhealthiest foods a person could eat. You must strictly avoid fiber in order to achieve a reduction in the symptoms of your autoimmune disease, and those fortunate enough to be free from these horrible diseases are more likely to keep themselves healthy by avoiding it.

The information in this book goes contrary to many of the common understandings in our society—hence the name **Absolute Truth Exposed**. I present scientific information because I expect you to have sincere doubts about what you are about to read. This is more of a science book than a health, diet, or nutrition book.

I have included Chapter 6 about brainwashing because I hope you will come to realize that we have all been given a monumental "snow job" in almost every area of our daily lives. Unfortunately, this chapter contains the bad news as well—reversing brainwashing is almost impossible. Hopefully, you will be one of the fortunate new members to join the **Absolute Truth Exposed** team. I wish you luck, and I hope you enjoy the rest of the book.

Absolute Truth Exposed

Volume 1

Chapter 1

Absolute Scientific Proof Carbohydrates Are Pathogenic
(Disease Causing)

Chapter 1

This book will prove that eating **red** meat and natural animal fats while restricting carbohydrates is not only healthy but will prevent and cure many diseases.

Unhealthy High-Carbohydrate Foods

Grains	Cereal	Bagels
Legumes	Fruit	Bread
Pasta	Potatoes	Yams
Sugar	Ice Cream	Candy

Carbohydrates Cause Nearly All Age-Related Diseases

Age-related diseases are thought to be unavoidable. Many people consider it normal to get one or more of these diseases as they age. They rationalize that they are simply unlucky or that others have "better genes," neither of which is true. Their health problems are most likely caused by their belief in the many popular myths and distortions about nutrition. They most likely got hooked by the low-fat, high-carbohydrate diet craze and are now suffering as a result. Heredity is used as the most common excuse instead of identifying the real culprit—carbohydrates. People flippantly say, "It runs in my family," or "My mother also had diabetes," or "My father also had high blood pressure and heart disease." I describe age-related diseases as *Excessive Carbohydrate Consumption Syndrome.*

The scientific evidence is clear. Carbohydrates are sinister, sly foods that have been getting away with murder. Carbohydrates have powerful allies. Their promoters grow, manufacture, and market thousands of different products made from fruits, grains, and starchy-vegetables. The supermarket space allotted to manufactured carbohydrate foods is about 80% of the store, yet the scientific minimum requirement for carbohydrates in the diet

Proof Carbohydrates Are Pathogenic

is zero. Carbohydrates are not essential elements for health. In fact, optimal health lies in keeping the amount of carbohydrates in the diet to a minimum. The body can maintain a perfectly normal level of blood glucose by making glucose from proteins (amino acids) and fatty acids (glycerol) in a process known as *gluconeogenesis*. Eating carbohydrates is not required to maintain a perfect level of glucose in the blood. The figure in the following reference book shows the entire process in graphic form. The process is reversible and can go either way in order for the body to maintain the proper blood glucose level without eating any carbohydrates.

> "Both triglyceride molecules and protein molecules may be broken down and converted to glucose in the liver. The process by which new glucose is formed from noncarbohydrate sources is called gluconeogenesis. In the process of gluconeogenesis, glucose is formed from lactic acid, certain amino acids, and glycerol portion of triglyceride molecules (Fig. 25.10)." *Principles of Anatomy and Physiology, Seventh Edition* by Gerard J. Tortora and Sandra Reynolds Grabowski, page 835.

The supermarket departments which contain the healthy essential proteins and essential fats are the fresh meat, fresh fish, seafood, and dairy. Everything else in the store is very high in carbohydrates, which turn to glucose, hype the metabolism, and trigger the release of disease-causing hormones, such as insulin, cortisol, and adrenaline.

We are told that a high metabolism is good for weight loss and good health. High energy drinks with caffeine and herbal additives that spike the metabolism are very popular, and many health foods contain high levels of calories from carbohydrates. These foods and drinks are not healthy. The advice is backward.

Chapter 1

A low metabolism is ideal for long life and good health; a high metabolism excites hormones in the body that eventually cause age-related diseases. A low metabolism is analogous to a diesel engine, which is known for longevity and high mileage without a breakdown. Diesel fuel is oil that the engine uses for energy and is similar to fats in the diet. A high metabolism is analogous to a nitro-methane drag racer, which gives a tremendous burst of energy but explodes the engine after a few races. The nitro-methane fuel is fast burning, similar to sugar in the diet.

USDA 2005 Food Guide Pyramid
Source: United States Department of Agriculture
Carbohydrates: 50% to 60% depending on the individual

The new 2005 *USDA Food Guide Pyramid* above adds more confusion with its many different pyramids from which you must make a selection. This system will cause the current epidemics of obesity, cancer, diabetes, and heart disease to become worse.

The pathogenic effects of carbohydrates are slow but sure. The label, *20-Year Rule,* was given to describe the length of time

Proof Carbohydrates Are Pathogenic

between the start of the high-carbohydrate diet and the onset of disease. The separate diseases, severity, and time needed to develop these diseases are directly related to the percentage of carbohydrates in the diet. In the advanced stage, several different diseases are prevalent in the sufferer before death occurs.

> **Carbohydrates are not essential in the human diet. The *Food Guide Pyramid* is scientifically wrong.**

William Banting (1797 - 1878) was a formerly obese undertaker in London, England, who wrote *Letter of Corpulence,* a 22-page booklet, in 1859, in which he explained how a fairly low-carbohydrate diet greatly restored his failing health in a very short time. His success was based on his persistence in trying every diet his doctors suggested until he stumbled upon the absolute truth regarding the connection between diet and health. To his dismay, the information was instantly rejected by the majority of physicians and citizens alike.

Carbohydrates displace essential protein and essential fats in the diet and create a double health reversal. The carbohydrates themselves cause disease, and the deficiency of protein and fats contributes to or causes other diseases.

Chapter 1

The effects of excess carbohydrate consumption generally begin to appear in one of two ways.

High-Carbohydrate Diet Causes

- Body fat accumulation which leads to obesity, diabetes, heart disease, cancer, gallbladder disease, degenerative bone diseases, cataracts, and many others.
- Damage to the intestinal tract which leads to leaky gut syndrome, inflammatory bowel diseases, and a medical textbook full of autoimmune diseases. These illnesses generally make the sufferer underweight and deficient in vitamins and minerals caused by poor digestion.

We can see that the high-carbohydrate diet makes some people fat and some people wretchedly thin depending on how the carbohydrates damage the body. Excess body fat is indeed unhealthy, but the assumption that thin people are healthy is not true. Studies have shown that people who live the longest have a little more body fat than what is considered to be ideal.

The primary high-carbohydrate foods to avoid are sugars, honey, flour, grains, legumes, fruit, milk, and starchy-vegetables. Do not eat honey. Honey is pure carbohydrate sugar consisting of fructose, glucose, sucrose, maltose, isomaltose, maltulose, turanose, kojibiose, erlose, theanderose, and panose. It is very confusing why people worship honey. Honey in the diet is the same as white sugar. People who think honey is healthy are wrong. They claim it is acceptable because ancient cavemen may have eaten it, and we somehow became adapted to honey as a healthy food. That is an incorrect conclusion. Ancient Paleo man may have eaten it once in a lifetime but probably not. He didn't want to be stung by bees any more than we do, and he didn't have any protective netting.

Proof Carbohydrates Are Pathogenic

Whole grains cause disease in both humans and animals. Whole grain breads and bagels are not healthy food as people have been led to believe. All grains have a very high level of omega-6 fatty acids, which have been scientifically proven to be pro-inflammatory. Grains are poor sources of protein because they do not contain all of the essential amino acids. Grains are the most allergenic of any food group. Grains contain pathogenic yeasts and fungi, which produce mycotoxins. (Myco is from the Greek word *mykes* or *mukos* meaning *fungus*. Toxin is from the Latin word *toxicum* meaning *poison*.) Fungus is the primary pathogenic organism (including molds and yeasts) which infects the body, particularly the colon. The toxins are secondary metabolites produced by the fungus. Grains are saturated with fungus. You can watch the fungus grow rapidly by simply wetting some whole grain in a bag or dish and leaving it on the counter.

The primary cause of autoimmune diseases is the consumption of grains. Multiple sclerosis, lupus, and rheumatoid arthritis have been rare in populations that did not consume grains.

The omega-6 fatty acids in grains have been scientifically proven to be pro-inflammatory.

Grain fed to feedlot steers makes them fat and causes intestinal diseases. The feedlot diet given to steers is almost identical to the *USDA Food Guide Pyramid* as both diets are very high in grains. The feedlot operator is deliberately making the steers fat because fatty beef is given a higher grade, receives the best price, and has the best flavor. The time in the feedlot is short. The steers are sent to slaughter prior to developing any serious health problems. People get fat and develop disease for the very same reasons. Grains are worse for humans because we are

Chapter 1

omnivores. Steers are herbivores, but the grains still make them fat and give them diseases.

People in primitive cultures who ate a diet consisting primarily of meat from the hunt, lived in relatively good health. Those who switched to a grain-based diet suffered poor health, diseases, and a smaller stature.

The design of our digestive organs and digestive enzymes demonstrates that mankind is basically a carnivorous (meat eating) species with the ability to digest carbohydrates. Our digestive system is similar to those of carnivores such as lions and dogs. We can digest fruits and vegetables, but they do not provide the health benefits as claimed. The essential macronutrients required in the human diet are protein and fats; fruits and vegetables contain no protein or fats. The damage to health by the consumption of these carbohydrates is in proportion to the quantity ingested.

I observed two shoppers in the fresh fruit department of the supermarket. The older woman was bent over with a hump on her back from severe osteoporosis. She was literally filling her cart with a wide variety of fruit. The second woman was very young, perhaps in her early twenties. She stood tall and straight as she also filled her cart with an assortment of fruit. Neither of them got the connection. The collagen matrix in bone that provides the strength is made of protein. Fruit has no protein—absolutely none.

**Prepare Yourself!
Absolute Truth will clash with your convictions.**

Fruit is not the healthy food many claim it to be. Fruit is mostly fructose sugar with some vitamins, minerals, and other nutrients.

Proof Carbohydrates Are Pathogenic

Those vitamins and nutrients are easily obtained from meat and non-starchy vegetables without the fructose. The body processes fructose from fruit in the same way as it processes fructose from soft drinks. There is no difference. Fructose is fructose no matter what the source. Fructose causes insulin resistance as scientific tests have proved. Fructose is highly addictive, and most people simply refuse to give up fruit no matter how sick they become. This is identical to lung cancer patients who continue to smoke cigarettes.

The claim that fruit and whole grains are healthy foods will be viewed by science in future centuries as the biggest lie and fraud in the history of the world. These future scientists will certainly question why present-day organizations of professional nutritionists and physicians could have believed such claims when contrary scientific evidence was clearly available.

Manufactured baby foods are very high in carbohydrates. The products labeled as containing meat usually contain a high level of carbohydrates as well. All carbohydrates are turned to glucose (blood sugar) during digestion. The blood sugar zings the baby's metabolism, and the result is a hyper-metabolism, which may be the root cause behind the epidemic rise in childhood attention deficit hyperactivity disorder (ADHD). Unfortunately, official health organizations don't have a clue about the cause or cure for these problems. Mother's breast milk contains 58% fat. Baby food on the supermarket shelf contains only 10% fat and 70% carbohydrates. Some of the fruit and vegetable baby foods contain no fat and 100% carbohydrates. These manufactured baby foods are creating either an epidemic of fat children or wretchedly thin walking sticks. Few children in our modern society are strong, muscular, and robust like ancient North American Indian and Eskimo children who ate a 100% meat diet.

Chapter 1

Other infants start on the road to insulin resistance when they are fed the standard high-carbohydrate diet. This has been proven to be the root cause of childhood and adult obesity, diabetes, allergies, and heart disease. The high levels of carbohydrates in infant foods are setting the child up for a lifetime of health problems.

Carbohydrates Trigger Disease-Causing Hormones

The hormones involved in the carbohydrate—disease loop are not the sex hormones but rather metabolic hormones. The process starts when carbohydrates are eaten in the form of sugars, such as sucrose, fructose, lactose, and others. Simple carbohydrates are molecules made by chains of glucose that are short. Longer glucose chains form carbohydrates that are classified as complex. The body breaks the chains apart until individual molecules of glucose are released into the bloodstream. Then the problems start. The body is very sensitive to the amount of glucose in the blood, commonly called *blood sugar*.

Two different types of cells require glucose because they lack mitochondria (powerhouse of the cell) and cannot use fatty acids for fuel. These are in a small part of the brain called the *midbrain*, which is about 1 inch (25 mm) long. Red blood cells are the second group. The lack of glucose (hypoglycemia) as energy for the brain can cause symptoms ranging from headache, mild confusion and abnormal behavior, to loss of consciousness, seizure, coma, and death. The body can maintain an ideal level of glucose by creating it in the liver from amino acids derived from protein or triglyceride fatty acids in a process called *gluconeogenesis*. This is how the low-carbohydrate diet results in a perfectly-controlled and stable blood glucose level. On the other hand, the high-carbohydrate diet results in the body's

Proof Carbohydrates Are Pathogenic

constant attempt to prevent blood glucose swings both to the low-side (hypoglycemia) or the high-side (hyperglycemia). This control is regulated by the hormone insulin. The glucose level is reduced by insulin. The insulin forces glucose into cells to produce energy or forces it to be converted into fat for storage. The adrenaline hormone raises the glucose level as an emergency backup system to prevent hypoglycemia.

Hypoglycemia is a train whistle's signal that the diabetes train is coming down the track. The diabetes engine is powered by carbohydrates and gaining speed. Nibbling complex carbohydrates throughout the day to control the blood sugar swings will do nothing more than slow the train for a year or two. The diabetes train can be stopped dead on the tracks only by avoiding all carbohydrates. The condition of uncontrolled blood sugar swings, called *diabetes mellitus* or *Type 2 diabetes*, has become epidemic in all English-speaking countries. It is now epidemic but will soon become a catastrophe.

Supermarkets offer fruit 365 days of the year. Obesity and diabetes are epidemic.

Younger people appear to handle carbohydrates without a problem. The younger body readily accepts the glucose with only a small insulin response and turns the glucose into energy. This constant bombardment of glucose causes the cells to become resistant to insulin. Ever-increasing levels of insulin are necessary to maintain a normal blood glucose level. As the cells become resistant, the insulin assists in the conversion of the extra glucose into triglycerides. The rise in the triglyceride level in the blood signals the body to store the excess as body fat. Carbohydrates cause obesity—not eating fat. Fruit juice in the diet easily becomes saturated fat on the midsection. Most people

Chapter 1

simply do not realize that the body converts carbohydrates to saturated body fat. The high-carbohydrate diet is a natural killer for many reasons.

Insulin Is a Disease-Causing Hormone

Insulin is a hormone made by the beta cells in the islets of Langerhans in the pancreas. Body cells require insulin in order to use blood glucose.

A high level of blood insulin causes many unhealthy body reactions which eventually lead to all types of diseases. Insulin turns glucose from the excessive consumption of carbohydrates into body fat and deposits it in the arteries and organs which results in arterial diseases, heart disease, strokes, blood clots, and other diseases. High blood glucose requires an ever-increasing amount of insulin production. Because the pancreas becomes fatigued after many years of excessive insulin production, diseases appear to be age related. Glucose rises uncontrollably when insulin production drops which results in diseases of the eyes, kidneys, blood vessels, and nerves.

Carbohydrates drive insulin production, and high insulin levels lead to coronary heart disease (CHD). Many heart attack patients first learn they are diabetic in the hospital emergency room, but they may not be told about the close relationship between their two conditions. Blood insulin reaches high levels and remains high as a person progresses from hypoglycemia to Type II diabetes. Insulin production eventually collapses in the late stage of Type II diabetes. Insulin is a very strong anabolic hormone which pushes blood glucose into cells and small, dense LDL molecules into the artery walls to begin the process of atherosclerosis. Animal research with insulin proved many years ago that the artery will plug with atherosclerosis just

Proof Carbohydrates Are Pathogenic

downstream from the point of injection. Insulin turns blood glucose into triglycerides and stores them as body fat. Insulin causes glucose to be chemically combined with cell membranes, LDL lipoprotein, blood hemoglobin, and other cells in a process called *glycation*, which I will discuss in more detail later. Insulin stimulates the growth of artery smooth muscle cells and causes arteriosclerosis (hard, thick arteries) and hypertension (high blood pressure). Insulin stimulates the growth of fibrous tissue, assisting in the formation of plaque in the arteries. Insulin stimulates the growth of fibrinogen that thickens the blood and forms blood clots. The entire process happens rather suddenly as insulin resistance peaks and insulin levels skyrocket. I have described the sudden appearance of heart disease as the *Rieske Instant Atherosclerosis Cycle (IAC)*.

Eating a low-carbohydrate diet brings unhealthy body processes caused by insulin to a screeching halt. All humans react in the same way. People are not different from one another as the myth implies. The variable among people is the amount of damage that carbohydrates and insulin have already done to the body. This damage is incorrectly diagnosed as genetic.

The scientific proof that insulin causes many age-related diseases is beyond question, yet the professional medical practitioners rarely measure their patients' insulin levels. Many diabetics have never had an insulin measurement taken by their doctors. Professional heart disease organizations rarely mention insulin as a heart disease risk or promote a low-insulin diet as a method of preventing heart disease. The carbohydrate—glucose—insulin cycle is getting away with mass murder while these professional organizations are saying, "Don't eat red meat or saturated fat." A few brave individuals, such as the late Dr. Robert C. Atkins and Drs. Michael and Mary Dan Eades, have

Chapter 1

promoted the science of health while enduring a constant assault from professional health and medical organizations. Professional medical associations almost universally ignore mountains of scientific evidence in favor of unproven myths.

Carbohydrates cause the LDL cholesterol molecules to be the unhealthy, small, dense fraction. All LDL molecules are not the same. The high-fat, low-carbohydrates diet causes the LDL molecules to become the safe, large, fluffy, light-density fraction. Higher LDL blood levels on the low-carbohydrate diet do not present the same cardiovascular heart disease risk as do LDL levels on the *USDA Food Guide Pyramid* diet of 60% carbohydrates.

High Insulin Increases Cancer Risks

Carbohydrates drive blood insulin production and cause cancer. There are strong associations between a high-carbohydrate diet and many diseases that present a secondary cancer risk. Cancer risks are greatly increased with diabetes and inflammatory bowel diseases. High levels of glucose and insulin cause poor circulation and many other unhealthy conditions that can lead to cancer.

The only way to prevent diseases caused by insulin spikes and plunges is to eat a low-carbohydrate diet. Many primitive societies lived on a natural diet that contained very few carbohydrates, and studies of the health of these societies proved that diabetes and all the diseases of consequence did not exist. Eskimos of the far north prior to the introduction of white-man's food are great examples. Primitive people in Egypt ate a natural diet high in carbohydrate content and suffered severely from cancer, diabetes, heart disease, degenerative disc disease, and many autoimmune diseases.

Proof Carbohydrates Are Pathogenic

The bad effects of insulin do not end here. High-insulin spikes signal the body to release cortisol and adrenaline hormones that also contribute to disease.

Insulin causes eye diseases. Diabetics suffer from diabetic retinopathy, retinitis pigmentosa, macular degenerative, optic neuritis, and blindness. Cataracts are most likely caused by insulin and the high-carbohydrate diet as proposed by the *USDA Food Guide Pyramid*. The damage probably occurs many years before the cataract first appears. Cataracts are not genetic as claimed. They most likely are not caused by free radical attack as often claimed. Many diseases are said to be genetic, when in fact they occur because people cook and eat the same foods as their parents.

High levels of blood glucose in the presence of hyperinsulinemia cause polysaccharide chains to attach to serum proteins, artery proteins, and LDL molecules in a process called *glycation*. The glycated protein and LDL form a fatty deposit called *atherotic plaque*. The serum molecules and the fatty deposits become sticky. White blood cells called *macrophages* ingest the glycated proteins and glycated LDL. The macrophages swell to form foam cells that attach to the sticky artery walls, thereby narrowing the artery and restricting blood flow. These conditions are caused by a diet high in carbohydrates. The process occurs rather suddenly in people who have hyperglycemia (high blood sugar), hypoglycemia (low blood sugar), or diabetes. Clear arteries can become plugged and cause a heart attack in as little as one year. This is commonly seen following artery bypass surgery because the replaced artery can become plugged within a year.

Chapter 1

High Insulin May Cause Alzheimer's Disease

The main cause of Alzheimer's disease is still very much an unresolved debate, but insulin is a prime suspect. High levels of insulin cause arteries to become plugged in places like the heart, kidneys, legs, and neck. Researchers have not focused their attention on the small capillaries, but trouble must surely be brewing there as well. Swelling of the feet and legs (edema) in diabetics and others could very well be caused by restricted capillaries. This is the possible connection to Alzheimer's disease because we know the brain tends to shrink with age and the progression of the disease.

Studies have shown that mice with the mouse model of Alzheimer's disease improve on a high-fat, low-carbohydrate diet (ketogenic) as recommended in this book. A high-carbohydrate, low-fat diet as outlined on the *USDA Food Guide Pyramid* increases Alzheimer's symptoms. The current epidemic of Alzheimer's disease did not come as a surprise to me. We can expect the trend to continue as government health departments, professional nutritionists, and physicians continue to pound the low-fat, high-carbohydrate dogma into the brains of the people in all English-speaking countries.

High Insulin Increases Alzheimer's Disease in Mice Studies

"A ketogenic diet reduces amyloid beta 40 and 42 in a mouse model of Alzheimer's disease." Ingrid Van der Auwera, Stefaan Wera, Fred Van Leuven, and Samuel T. Henderson. *Nutrition & Metabolism*, BioMed Central, United Kingdom.

> "Mice with the mouse model of Alzheimer's disease show improvements in their condition when treated with a high-fat, low-carbohydrate diet. A report published today in the peer-reviewed, open access journal Nutrition and

Proof Carbohydrates Are Pathogenic

Metabolism, showed that a brain protein, amyloid-beta, which is an indicator of Alzheimer's disease, is reduced in mice on the so-called ketogenic diet."

"The report, by Samuel Henderson, from Accera, Inc, Colorado and colleagues from Belgium runs counter to previous studies suggesting a negative effect of fat on Alzheimer's disease."

High Insulin Causes Obesity

Obesity results from an abnormally high level of insulin due to insulin resistance. Insulin resistance is caused by the continual consumption of carbohydrates, such as whole grains, legumes, potatoes, honey, fruit, and sugar. The adipose tissue (body fat) increases by lipogenesis—the conversion of carbohydrates to fat. The body cannot burn this body fat for energy in the presence of elevated insulin. Glucagon is a hormone that has the opposite effect. It causes the liver to convert fats into fatty acids and ketone bodies (acetoacetic acid, beta-hydroxybutyric acid, and acetone) for conversion to energy by the body. Glucagon increases glucose by the conversions of body fat through lipolysis—the breakdown of triglycerides into free fatty acids within cells.

> **The body can quickly and easily turn the fructose in orange juice into saturated fat.**

All carbohydrates are digested and enter the blood as glucose, which in turn causes a surge in the insulin level. Since the body is very sensitive to the amount of glucose in the blood, the insulin quickly converts the glucose to triglycerides for transport to the adipose tissue for storage. The insulin drops the glucose level, which causes renewed hunger, and the vicious cycle continues.

Chapter 1

The low glucose level causes the obese to be lethargic. The lack of energy is a result of the insulin blocking the conversion of adipose tissue to energy. Carbohydrates cause obesity and laziness in those who have become insulin resistant. The only diet that allows the obese to lose weight and maintain the weight loss is the low-carbohydrate diet.

Cortisol Is a Disease-Causing Hormone

Cortisol is the major stress hormone of the natural glucocorticoid family. Cortisol regulates metabolism and provides resistance to stress. Glucocorticoids are made in the outside portion (the cortex) of the adrenal gland and are chemically classified as steroids. They increase the rate at which proteins are catabolized (broken down) into amino acids. The amino acids are removed from cells—primarily muscle fiber—and transported to the liver. Glucocorticoids cause amino acids to be synthesized into new proteins, such as enzymes or immune cells, and they also raise blood pressure by constricting vessels—a benefit in case of injury. They are also anti-inflammatory. All of this is well and good in a healthy individual with normal glucose and insulin levels. Unfortunately, high cortisol levels cause many unhealthy reactions.

Hyperglycemia results when a high ratio of carbohydrates to protein in the diet increases the glucose in the blood. The body then reacts by increasing insulin. However, the effect of the insulin is delayed in a person who is insulin resistant or glucose intolerant. This is often referred to as *metabolic syndrome*. The glucose is not reduced in a timely manner as in a healthy individual, and the body continues to increase insulin until the effect is excessive. The excess insulin causes the glucose level to drop which results in hypoglycemia. The body attempts to correct the low blood glucose level by stimulating the adrenal

Proof Carbohydrates Are Pathogenic

gland to secrete additional cortisol and adrenalin. The individual is on the glucose roller coaster of highs and lows—an early stage of diabetes. This chain of events occurs when a high-carbohydrate diet leads to harmful levels of cortisol.

Excess Cortisol Causes Many Problems

- Diminishes cellular utilization of glucose.
- Increases blood sugar levels.
- Decreases protein synthesis.
- Increases protein breakdown that can lead to muscle wasting.
- Causes demineralization of bone that can lead to osteoporosis.
- Interferes with skin regeneration and healing.
- Causes shrinking of lymphatic tissue.
- Diminishes lymphocyte numbers and functions.
- Lessens secretory antibody productions (SIgA). This immune system suppression may lead to increased susceptibility to allergies, infections, and degenerative disease.

The excessive consumption of carbohydrates leads to high insulin that in turn leads to high cortisol. The cortisol causes the body to extract high-tensile strength collagen protein fibers from bones, remove the mineral matrix by demineralization, and weaken connective tissue at the joints. The protein loss is accelerated by a low-protein diet. Cortisol literally causes people to pee away their bones. The result is a rapid onset of osteoporosis and degenerative disc disease in which the spine can lose as much as one inch (25 mm) in height in as little as one year. Bones fracture more easily, and the dreaded hip fracture is much more likely to occur. I have described this sudden onset of osteoporosis as the *Rieske Instant Osteoporosis Cycle (IOC)*. Women are told to drink milk and eat yogurt to get additional calcium. The promise that this will prevent bone loss is based on faulty logic. The lactose in the milk and yogurt along with the sugar and fruit added to yogurt only serve to increase the dietary carbohydrate load. The net result is harmful to the bones as many are discovering.

> **Cortisol literally dissolves your bones.
> Osteoporosis is not a calcium deficiency.**

A low level of blood glucose induces higher levels of cortisol. Therefore, a person with hypoglycemia generates unhealthy cortisol as the blood glucose level swings from above normal to below normal.

The catabolic state imposed by high-cortisol and adrenaline levels is opposite to the anabolic body-building state desired by athletes. Catabolism breaks down and destroys the body, which leads to a myriad of chronic diseases; many are deadly and others cause years of suffering. The disastrous effects of this sudden catabolic state cannot be overstated.

Catabolic State Causes Many Problems

- Osteoporosis bone loss, hip fracture, and degenerative disc disease.
- Weakened ligaments and tendons which reduce joint mobility and cause pain.
- Weak and torn muscles that occur at random, reduce strength, and cause pain.
- Osteoarthritis caused by destruction of the joint-lubricating protein structures.
- Weakening of the myocardial muscle which leads to chronic heart disease.
- Weakening of the valves of the heart which leads to problems such as mitral valve prolapse.

Proof Carbohydrates Are Pathogenic

- Weakening of the septal valve that closed at birth until it reopens between the heart chambers.
- Thinning of blood vessels which may lead to a hemorrhagic stroke or a ruptured aneurysm.

Cortisol is a counter-regulatory hormone. Increases in cortisol may decrease the effectiveness of other hormones such as testosterone and estrogen, thereby diminishing their health-protecting effects. All of this can be prevented by eating a high-protein, high-fat, low-carbohydrate diet.

Adrenaline Is a Disease-Causing Hormone

Adrenaline (epinephrine) is the "fight-or-flight" stress hormone. Epinephrine is a neurotransmitter secreted by the adrenal gland and is associated with sympathetic nervous system activity. It prolongs and intensifies the following effects of the sympathetic nervous system.

Excess Adrenaline Causes Many Problems

- Causes the pupils of the eyes to dilate.
- Increases the heart rate, force of contraction, and blood pressure.
- Constricts the blood vessels of nonessential organs such as the skin.
- Dilates blood vessels to increase blood flow to organs involved in exercise or fighting off danger, skeletal muscles, cardiac muscle, liver, and adipose tissue.
- Increases the rate and depth of breathing and dilates the bronchioles to allow faster movement of air in and out of the lungs.
- Raises blood sugar as the glycogen in the liver is converted to glucose.
- Slows down or even stops processes that are not essential for meeting the stress situation, such as

Chapter 1

muscular movements of the gastrointestinal tract and digestive secretions.

All of these effects are wonderful if one is being chased by a lion or attacked by an intruder in the home. However, these effects are unhealthy to a person sitting in an office, watching a football game, or simply going about his everyday life.

The last item on the above list is very disruptive to the intestinal tract and results in intestinal diseases because adrenaline slows down or stops digestive secretions and muscular movements of the gastrointestinal tract. People are advised to eat more high-fiber whole grains and high-fiber fruits and vegetables to overcome the constipation that results from this slowdown of the intestinal system, but this advice is backward. All of the high-fiber foods are also high-carbohydrate foods—just the foods one should avoid for stopping the surges in insulin and adrenaline that shut down the digestive processes. Eating fiber is counterproductive. The next chapter will explain in more detail why the information being disseminated about dietary fiber is false.

Hyperinsulinemia (high-insulin) and hypoglycemia increase adrenaline when no fight-or-flight stress situation exists; this creates unhealthy body changes. These changes include effects to the cardiovascular system that increase the risk of coronary heart disease. The low-fat, high-carbohydrate diet as outlined on the *USDA Food Guide Pyramid* causes disease because it promotes hypoglycemia, hyperinsulinemia, hypertriglyceridemia, and hyperadrenalemia. A prolonged elevation of adrenaline has the following effects on the cardiovascular system.

Proof Carbohydrates Are Pathogenic

Excess Insulin and Adrenaline Together Cause Many Problems

- Increase the production of blood cholesterol, especially the undesirable LDL.
- Decrease the body's ability to remove cholesterol.
- Increase the blood's tendency to clot.
- Increase the deposits of plaque on the walls of the arteries.

Adrenaline addiction is very common. Type-A personalities become addicted to their excessive activity by the stimulating effects adrenaline. People who are constantly angry, fearful, guilty, or worrisome arouse their adrenaline hormone even though they may be sitting around doing nothing. People who excessively participate in jogging, exercise, bodybuilding, aerobics, sports, skiing, biking, mountain climbing, car racing, or flying aerobic airplanes become addicted to the adrenaline rush. They describe the rush they get from their activity and feel depressed when they can't participate for some unexpected reason. James F. Fixx was addicted to running and wrote the famous jogger's book, *The Complete Book on Running*. He was a marathon runner and vegetarian on a diet of high carbohydrates and low protein. This combination was a perfect setup to arouse and maintain a high level of adrenaline. He died of a massive heart attack on his daily run—proving to the world that exercise does not prevent coronary heart disease. Fixx admitted in his book that his own research showed the athletes from his university alumni had a shorter life span than the "couch potato" students. This may have been due to the difference in adrenaline levels. Hyperglycemia and stress are a deadly combination.

Chapter 1

The only acceptable diet that prevents high-insulin, high-glucose, and abnormal adrenaline secretion is the high-protein, high-fat, low-carbohydrate diet.

Carbohydrates Decrease Human Growth Hormone

The consumption of excess carbohydrates raises the glucose level in the blood. In turn, the abnormally high blood glucose level stimulates the hypothalamus to secrete growth hormone releasing hormone (GHRH), which inhibits the release of human growth hormone (hGH). A drop in human growth hormone diminishes muscle growth and slows bone growth. The net result is an increase in body fat with poor muscle tone; degenerative disc disease; and weak bones, along with the destruction of hip; knee; and shoulder joints. This body structure is obvious in people who are obese.

The bodies of adolescent children on the high-carbohydrate diet can grow in either of two directions. Some develop a condition known as *insulin resistance* from the onslaught of glucose in the blood. They become fat and lethargic. They have low energy and are accused of being lazy. They sit in the school classroom and doodle on scratch paper. They don't talk much and are antisocial. They flop in front of the television and avoid physical activities.

The other extreme is hyperactive children who are not insulin resistant. The onslaught of glucose over-fuels the muscle cells of the body, and they become jumpy, overactive, and unable to sit still. When they are in school, the hyperactive students talk out of turn, disturb other students, and require constant discipline by the teacher. The schools insist that parents place these children on brain-altering drugs to keep them calm. The result is a hyper metabolism that may be the root cause behind the epidemic rise in childhood attention deficit hyperactivity disorder (ADHD).

Proof Carbohydrates Are Pathogenic

Unfortunately, official health organizations don't have a clue about the cause or cure for these problems. Children are falsely diagnosed as having a mental or emotional handicap when the real problem is their high-carbohydrate diet. They are actually carbohydrate addicts. Their school lunches give them another big shot of carbohydrates, and the hyperactivity is off and running again. Hyperactive children must be constantly on the move in order to burn up the excess carbohydrate fuel.

Childhood attention deficit hyperactivity disorder (ADHD) is caused by the high-carbohydrate diet.

Hyperactive children may grow taller than other children. Their thin frames and poor muscle tone make them appear as "walking sticks." They must stand twice in the same spot to cast a shadow. You see them everywhere, running, jumping, playing sports, and talking constantly. In both cases, the children are protein and fat deficient.

Children quickly become carbohydrate addicts. Young children—say two years of age—throw a fit if necessary to get candy and desserts. They scream, cry, pout, and physically attack the parents in order to get the glucose high. They can be seen embarrassing their mothers in the supermarket check-out lines as they throw temper tantrums and demand more candy. Older children who are carbohydrate addicts simply help themselves to the sweets at home and buy them at the store. These carbohydrate-addicted children begin the meal by eating all the sweets on their plate first. They gorge on the fruit first before turning to the whole grains. They love the starch in potatoes because it turns to glucose quickly. By this time, they have satisfied their glucose cravings for a few minutes and refuse to eat any meat. Consequently, they did not eat the "balanced diet"

Chapter 1

that may have been placed before them. They ate only carbohydrates. They are screaming for candy or dessert five minutes after finishing the meal. Blood sugar swings are making some of them insulin resistant. These children are society's future diabetics.

Proof Carbohydrates Are Pathogenic

Carbohydrate Death Curve

The long and short-term effects of eating carbohydrates are shown on the graph below. Degenerative diseases, such as heart disease, stroke, diabetes, cancer, inflammatory bowel diseases, rheumatoid arthritis, multiple sclerosis, psoriasis, lupus, osteoporosis, and many other diseases increase as the percentage of carbohydrates in the diet increases. Excellent long-term health is achieved on the left side of the curve where lower amounts of carbohydrates are consumed. Carbohydrates are not a dietary requirement in the human diet. This scientific physiological truth applies equally to all humans. The detrimental effects of high levels of carbohydrates accumulate over the years, producing the degenerative results that appear to be age related. Age is only a secondary factor and is evidenced by the fact that some older people do not have these diseases. The short-term climax occurs at the far right end of the curve, where a diet consisting of 100% carbohydrates will result in a quick and certain death within weeks.

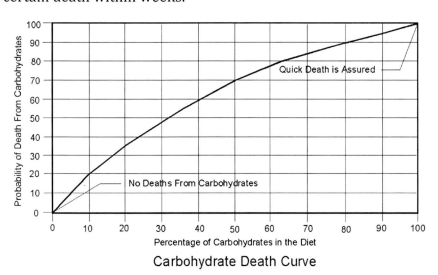

Carbohydrate Death Curve

Chapter 1

Recent Testimony of an Athlete

On December 5, 2009, Michel wrote from Sweden about his awesome improvements in health and athletic performance after only a few weeks on the low-carbohydrate diet presented here.

"Hi!

Just found your diet recommendations for athletes and I have to agree with everything you say!

I switched to a low-carb high-fat diet a couple of months ago and I am free from the problems I had. Discovered that I was gluten intolerant and could quit my Omeprazol medication and found out that a lot of other minor issues went away by just cutting out carbs. I'm an ambitious athlete aiming to try to reach elite, but because I got very sick in flus, colds and respiratory infections for nearly 2 months all hard training was lost. When I switched diet my health got better already after 2 weeks. I have also discovered that I have almost doubled my endurance just by not eating carbs at all - which is the opposite of what athletes have been told the world over. Now I can run and ride the bike for ever without hitting the wall, all I need is water. I haven't felt any lactic acid either, it's gone. My recovery time is faster than before. I run much better on fat than I did on carbohydrates. Athletes have been told to "carb load" which for me has been worthless, I mean 100g of sugar have been proved in studies to limit your immune system with 40% - not a good idea for an athlete who depends on his health to perform. Fat loading is the only healthy way to go!

Proof Carbohydrates Are Pathogenic

Your approach is very straight forward and just the stuff we need to change this sickening carbohydrate idiocy!

Best regards,

Michel"

Unhealthy High-Carbohydrate Foods

Grains	Cereal	Bagels
Legumes	Fruit	Bread
Pasta	Potatoes	Yams
Sugar	Ice Cream	Candy

Chapter 1

Ketosis Facts and Myths on the Low-Carbohydrate Diet

Most people experience a condition called *ketosis* when they suddenly go from a high-carbohydrate diet to a low-carbohydrate diet. This occurs when ketone molecules are circulating in the blood in a higher amount than on the previous high-carbohydrate diet. Contrary to the myths, distortions, and lies published by vegetarians and other supporters of a high-carbohydrate diet, ketosis is a normal physiological state that results from very normal and healthy body functions. Ketosis allows the body to function efficiently and live off of stored body fat when necessary. Ketones are not a poison as many medical and nutritional experts falsely claim. Ketones make the body run more efficiently and provide a backup fuel source for the brain. The three substances known as *ketone bodies* are acetoacetic acid, beta-hydroxybutyric acid, and acetone. Since acetone is a common household solvent found in fingernail polish remover, some misinformed people have an absolute hissy-fit when told that ketosis produces acetones in the blood. The body produces ketones as the preferred fuel for the heart, outer part of the kidneys, and most areas of the brain.

> **Ketosis is a healthy condition.**
> **Ketoacidosis is a deadly ailment.**

Ketoacidosis is a life-threatening condition commonly associated with Type 1 diabetes and insulin-dependent Type 2 diabetes. Ketoacidosis is not the same as normal dietary ketosis. The abnormally low level of insulin in the diabetic leads to a toxic build-up of blood glucose which causes excess urination, thirst, and dehydration. The glucose cannot enter the cells to produce energy in the absence of insulin; therefore, the body must break down an excessive amount of body fat and muscle tissues for

Proof Carbohydrates Are Pathogenic

energy. Ketoacidosis is an unhealthy condition in which the body has excessively high glucose and ketone bodies at the same time. The ketoacids lower the pH of the blood to an unhealthy level. The body is very sensitive to changes in blood pH. Ketoacidosis is never present in a non diabetic on a carbohydrate-restricted diet in which beneficial ketosis is achieved. Normal ketosis stabilizes blood glucose within a normal range and prevents the breakdown of healthy muscle tissue. Because the dietary restriction of carbohydrates prevents any buildup of excess glucose, the blood glucose level remains perfectly normal and stable—almost straight line. The body is powered normally by ketone bodies while we sleep. Ketosis was common and normal in all primitive people, such as the North American plains Indians and northern Eskimos, who lived on a high-protein, high-fat, and low-carbohydrate diet. Dietary fatty acids and ketone bodies are very healthy fuels for the body as opposed to glucose, which causes insulin resistance, diabetes, heart disease, and cancer.

Many people falsely believe they are unable to lose weight because their ketones drop to a lower level. The Ketostiks® are no longer showing the purple color that indicates the presence of ketones in the urine. The common assumption that ketosis must be present in order to lose weight is incorrect. Weight loss continues with or without the presence of ketones. The spillage of excess ketones into the urine disappears after the body becomes accustomed to using dietary fats for energy instead of glucose; this usually occurs within a few weeks of starting the diet. A person who always eats a high-carbohydrate meal is fueling his body with glucose and fructose obtained from carbohydrates. Because the cells have had very little experience burning fatty acids for energy, the body experiences a new condition on the low-carbohydrate diet. The more commonly-used glucose fuel is no longer available. When the pancreas

Chapter 1

drops the production of insulin, the body reacts by increasing the production of the glucagon hormone. The glucagon draws stored fat reserves in the form of triglycerides for the cells to use as the new energy source. But the cells are slow to react to this new fuel source, and the individual feels weak or lacks energy. The resistance to burning fatty acids for energy can vary greatly among individuals. Some people feel this weakness, but others pick right up and take off with the feeling of increased energy.

The liver begins to catabolize (break down) the extra fatty acids when they are not being utilized by the cells. However, since the liver does not have the enzyme necessary for complete catabolism of the fatty acids, ketone molecules are discharged into the blood. The strange taste in the mouth and mild breath odor indicates the presence of these ketones. Body cells can utilize the ketones for energy. Contrary to the myth that it must be powered only by glucose, the brain can utilize ketones; although some areas of the brain still require glucose, which the body makes from amino acids or fats. Ketones are the preferred energy source for the muscles of the heart.

Ketones are the preferred fuel for heart muscle.

The body begins to utilize the fatty acids for energy more efficiently after a few weeks on the low-carbohydrate diet and the ketones drop to a lower level. This does not mean a person is not losing weight. It means the body is becoming a more efficient fat burner.

The brain and heart can endure harsh starvation as can be seen in the pictures and movies of the rescue of German concentration camp inmates at the end of World War II. The poor victims were almost completely depleted of muscle mass because their bodies

Proof Carbohydrates Are Pathogenic

were using muscle protein as fuel to keep them alive. When the survivors were interviewed many years later, they demonstrated excellent mental capacity, and many lived to very old age. The heart muscle eventually gives out during starvation, but that does not occur until the person is too weak and frail to stand or walk. The brain and heart survive surprisingly well during starvation. The observation of German concentration camp inmates proves that a low-carbohydrate diet has no detrimental effect on the brain or heart as falsely claimed by many professional nutritionists.

The scientific and practical amount of carbohydrates needed in the diet is zero—none. Dietary carbohydrates are not a requirement in any body function. In fact, it appears that the lower the level of carbohydrates in the diet, the better the long-term health. Age-related degenerative diseases are caused by high levels of carbohydrates in the diet. One can be on a very low or zero-carbohydrate diet indefinitely as studies of many primitive societies have shown. I know I am repeating this information, but repetition is necessary in order to overcome the detrimental effects of long-term brainwashing.

When I was on the Atkins' Induction Phase, I ate fewer than 20 grams of carbohydrates per day for many months. Ketosis had long since disappeared. I went for a blood test at a Quest Diagnostics® walk-in facility after the required 12-hour fast. The results showed a perfectly normal blood glucose level in the midrange. It could not have been better. Carbohydrates are not required to maintain a normal blood glucose level and not required in the diet for brain fuel. Don't believe the brainwashing statements that claim otherwise. The body can make all the glucose it requires from protein and fats in the diet or in the body itself in order to keep the glucose at a healthy level.

Chapter 1

Dealing with Low-Carbohydrate Ketosis or Hypoglycemic Symptoms

The low-carbohydrate ketosis diet typically produces a mild ache in the forehead. This is especially true for those new to the diet. In most cases, the feeling is unlikely to be hypoglycemia even though people think that it is. Simply ignoring the mild headache is one possibility. Here is some other helpful information:

- **Get a Free Blood Glucose Meter.** Buy a glucose meter that offers a full-cost rebate. The best brand and model to buy depends on the number of test strips included because replacement strips are expensive. It is simple to use with a small lancet and a lancet holder that looks like a pencil. Some lancets should also be included. The blood glucose level is most likely to be highest for a non diabetic (100 to 110 mg/dl) upon awaking in the morning because the liver produces glucose during the night. Readings less than 100 during the day are normal on the low-carbohydrate diet.

- **Hypoglycemia.** The low-carbohydrate headache is not hypoglycemia unless the blood glucose level is below 70 mg/dl.

- **Snack.** Eat a good high-fat meat for a snack. Baked or fried chicken thighs are great. The body can convert 58% of protein into blood glucose as needed without raising insulin levels. Meat in a snack prevents hypoglycemia naturally without eating sweets. Canned salmon is the perfect food because it also has a high level of essential DHA and EPA omega-3 fatty acids. Make sure the salmon includes water and salt only. Do not buy salmon with soybean or other bad omega-6 vegetable oils. Do not eat tuna fish as is commonly recommended because it does not have any omega-3 fats. Hard cheeses are also excellent snack foods, but

Proof Carbohydrates Are Pathogenic

the salt content can be a concern. Swiss cheese usually has the lowest sodium content.

- **Minerals, Vitamins, and Supplements.** Optimal health requires an abundance of vitamins, minerals, and supplements. The full array of colloidal minerals is very important.

- **Protein and L-Glutamine Drink.** Prepare a drink made with whey amino acid protein powder that is enriched with extra glutamine amino acid. The protein powder consists of a full complement of amino acid isolates that heal the body and require no digestion. Prepare the drink by blending 8 to 16 oz of reverse osmosis (R.O. with UV lamp) water or unsweetened, low-sodium tomato juice with 1 heaping teaspoon (12 gm) of whey protein powder plus 1 rounded teaspoon (8 gm) of glutamine amino acid powder. Stirring vigorously with the teaspoon is sufficient. The whey protein must be specified on the carton as isolates from cross flow microfiltration and ion-exchange, ultra-filtered concentrate, low molecular weight, and partially hydrolyzed whey protein peptides rich in branched chain amino acids and glutamine peptides. The low-carbohydrate type at 1 gm per scoop or less is best, but it should not be more than 4-5 gm of carbohydrates per scoop. Do not substitute protein from soy, egg, casein, or any other source. Sugar or any other sweetener is unacceptable. Use the "natural flavor" without additives. This amino acid drink can be enjoyed anytime with or without a meal. Amino acids are foods that build and maintain the body. **Refrigerate whey protein powder and discard if it is old. Whey protein powder can cause some gas and an unusually full feeling. Discontinue the whey protein powder if the reactions are unpleasant. Continue to take the glutamine powder.**

Chapter 1

This combination of amino acids has been shown to provide the following healing properties:

- Provides pain killing effects by healing the nervous system.
- Allows absorption of body building amino acids without requiring digestion.
- Stimulates insulin-like growth factor 1 (IGF-1), which functions similarly to insulin and enhances protein synthesis and healing.
- Fights infections by stimulating the immune system. All immune cells are made from poly-peptides of amino acids.
- Provides bone growth of protein collagen and strengthens bones. Poor digestion has been shown to cause osteoporosis and degenerative bone disease.
- Provides all of the amino acids required to heal and grow ligaments, tendons, joints, muscles, intestinal tract, heart muscle, and all other organs of the body.
- Prevents hypoglycemia (low blood sugar) symptoms in people with hypoglycemia or diabetes.

The three most nutritionally-important omega-3 fatty acids are alpha-linolenic fatty acid (ALA), eicosapentaenoic acid (EPA), and docosahexaenoic fatty acid (DHA). Alpha-linolenic fatty acid is one of two fatty acids traditionally classified as *essential*. The other fatty acid traditionally viewed as essential is an omega-6 fat called *linoleic acid*. These fatty acids have traditionally been classified as essential because (1) the body is unable to manufacture them and (2) they play a fundamental role in several physiological functions. As a result, we must be sure our diets contain sufficient amounts of both alpha-linolenic acid and linoleic acid.

Absolute Truth Exposed

Volume 1

Chapter 2

Absolute Scientific Proof Dietary Fiber Is Unhealthy

Chapter 2

The History of the Dietary Fiber Theory

The reason that most doctors, nutritionists, the *USDA Food Guide Pyramid*, and the official *USFDA Recommended Daily Allowance* (RDA) recommend a high-fiber diet is based on the faulty logic of Dr. Dennis Burkitt, a British surgeon working in Africa more than half a century ago. Dr. Burkitt's theory that barley bread prevented inflammatory bowel diseases has led to the current dietary fiber theory. Dr. Burkitt's fiber theory, developed in the late 1960's, was based on an impromptu hunch that fiber in the diet accounted for the difference in digestion system health between his African patients and white Englishmen living in Africa and England. His opinion was not based on any scientific studies; he did not perform controlled tests; and he did not isolate the groups with only fiber as the variable. His assumption that fiber in the diet made his African patients healthier was wrong.

Unhealthy High-Fiber Foods

Bran Cereal	Oats	Rice	Soybeans
Legumes	Split Peas	Lentils	Apples
Broccoli	Potatoes	Grapefruit	Pears

The reason Dr. Burkitt concentrated on fiber instead of the many other nutritional differences is unknown. He was familiar with the intestinal problems in the peasants and common people of England in that day. White Englishmen living in Africa ate the same imported refined foods as their own countrymen in England. When the two diets are compared, it is very obvious why the Africans had better health.

- The English ate large quantities of sugar. The Africans ate none.

Proof Dietary Fiber Is Unhealthy

- The English ate large quantities of refined flour. The Africans ate none.
- The English ate large quantities of breakfast cereals made from grain. The Africans ate none.
- The English ate large quantities of potatoes. The Africans ate none.
- The English ate salt-cured meat when meat was eaten. The Africans ate fresh meat, not cured meat.
- The English ate very little fresh meat or raw meat. The Africans ate generous quantities of both.
- The English were most likely protein deficient. The Africans had a generous supply of protein from meat.
- The English diet was deficient in omega-3 fatty acids. The Africans had omega-3 fatty acids in the fresh meat.
- The English diet was deficient in vitamins and minerals in processed foods. The Africans' fresh food provided an abundance of vitamins and minerals.
- The English diet caused rotten teeth and weak bones. The Africans had awesome teeth and bones.
- The English drank large quantities of sugar-sweetened soft drinks. The Africans drank none.
- The English ate a significant amount of honey. The Africans ate none.
- The English ate molasses and maple syrup. The Africans ate none.
- The English ate a significant amount of canned fruit. The Africans ate very little fresh fruit and no canned fruit.

The Africans were simply showing the benefits of eating a relatively low-carbohydrate diet consisting of fresh meats with animal fats and a few vegetables instead of the English diet that was high in carbohydrates, such as sugar, flour, cereals, honey, syrup, soft drinks, and potatoes. The barley bread was a very small part of the Africans' diet. Vegetarians have told lies about the true diet of the Africans, who were mainly meat-eaters—not vegetarians. The Africans raised domestic cattle, sheep, and

Chapter 2

goats; and they hunted and ate an abundance of wild animals. Fresh meat was very plentiful to the Africans, and they consumed it in great quantities. In addition, the Africans consumed all of the fat from the animal because it was considered to be the best part. The English people tended to cut off and discard the animal fat. The English also ate very little meat, but what they did eat tended to be processed and heavily salted. The diet of the Englishmen was protein deficient, but the Africans' diet was rich in protein and fat from fresh animal meat. The protein deficiency and high-carbohydrate diet of the English were the primary causes for the intestinal diseases and colon cancer observed by Dr. Burkitt.

Fiber scavenges minerals from the food and removes them from the body.

Why did medical and nutritional professionals immediately accept Dr. Burkitt's fiber theory while overwhelming evidence against it was ignored? The reason is intuitively obvious. Meat has no fiber. Vegetarians will promote any and every lie possible to stop you from eating animals. The dietary fiber theory is religion based.

Dr. Weston A. Price, DDS, visited the African Masai tribe in 1935 because they were noted for having excellent health. The average male stood well over six feet tall (1829 mm) and had perfect teeth, strong bones, and no intestinal diseases. The tall women gave birth easily to healthy, robust babies. The Masai lived entirely off of their cattle. They herded cows, drank their blood and milk, and ate the meat with all of the fat. The Masai refused to eat grains or grass seeds, claiming the grains and seeds were cows' food and not suitable for human consumption. The amount of fiber in the diet of the Masai was zero—absolutely none. The

Proof Dietary Fiber Is Unhealthy

milk, blood, meat, and fat eaten by the Masai have no fiber. The Masai proved that fiber is not a dietary requirement for perfect digestion and cancer-free colon health. According to Dr. Burkett's fiber theory, the Masai should have had an epidemic of intestinal diseases and colon cancer. They had none. Dr. Burkett's fiber theory is pure nonsense.

> **Dietary fiber is very bad for people who have inflammatory bowel diseases.**

Dietary fiber is not required in order to have perfect intestinal regularity. In fact, fiber actually encourages pathogenic bacteria growth in the colon. The toxic byproducts produced by the pathogenic bacteria are some of the principal causes of inflammatory bowel diseases. Fiber does not prevent or cure inflammatory bowel diseases but actually makes them worse. Studies of many other primitive societies have proved low-fiber diets prevent intestinal diseases and cancer as presented by Weston A. Price, DDS, in *Nutrition and Physical Degeneration.* In the summer of 1933, Primitive Indians were found living in the forest of Northern Canada, isolated from all civilization. The region was "inside" or east of the Rocky Mountain range in Northern British Columbia and the Yukon Territory. Access to the area was extremely difficult. On page 74, Dr. Price says, "The diet of these Indians is almost entirely limited to the wild animals of the chase. The rigorous winters reach seventy degrees below zero." (-57 degrees C.) He says on page 75, "Many of the women had never seen a white woman until they saw Mrs. Price." Dr. Price took a picture of a typical primitive family. (See page 76 of his book.) He says, "This typical family of forest Indians of Northern Canada presents a picture of superb health. They live amidst an abundance of food in the form of wild animal life in the shelter of the big timber." These Indians ate a diet of

Chapter 2

100% meat with the animal fat. The Indians ate the internal organs as well as the flesh and fat. Because the adrenal glands are the richest source of vitamin C in all animal or plant tissues, they never had any signs of scurvy (a disease caused by a vitamin C deficiency). They caught grizzly bear with baited pitfalls—a pit with a weak cover into which the animal fell but could not get out. Indians who lived closer to civilization and obtained white man's food suffered greatly with severely degraded health. These civilized Indians had access to carbohydrates and fiber in the diet. The primitive Indians had none.

Living a primitive lifestyle is not the single key to awesome health as many books claim. Primitive people who switched from an animal-hunting lifestyle to gathering or farming always had an immediate plunge in health, body stature, and height. Grains have been shown to be the most detrimental. Primitive societies who switched from a diet based primarily on meat from the hunt to a diet primarily of grains from farming were stricken almost out of existence.

During their first of two expeditions in 1906, Arctic explorers Vilhjalmur Stefansson and Karsten Anderson went to the Mackinzey River Delta in the Northwest Territories of Canada, where the river flows north into the Beaufort Sea of the Arctic Ocean. This is north of the present villages of Fort McPherson and Inuvik. (See Stefansson's book, *MY LIFE WITH THE ESKIMO.*) The Arctic Eskimos ate an all-meat diet with zero fiber. Their primary food was boiled salmon caught from the river. They ate salmon for breakfast, lunch, dinner, and snacks. Rotting or putrefied raw salmon and the crunchy heads were two of their favorite delicacies, which caused them no ill effects whatsoever. The Eskimos ate some animals that were hunted for the hides, such as caribou, bear, fox, and wolf. The small amounts of berries

Proof Dietary Fiber Is Unhealthy

they ate in season were of no consequence. The Eskimos had very healthy digestive systems with no cancer of any kind in the entire population. The digestive health of the Eskimos was much better than that of Dr. Burkitt's African patients who ate the higher-fiber barley bread diet. Mr. Stefansson could not detect one tooth cavity in all of the people or in any of the 2000 skulls he examined. Their skeletal health was excellent even though they ate no high-calcium foods like modern dairy products.

The low-fiber diet that the North American Plains Indians ate for centuries provided excellent health. They did not suffer from inflammatory bowel diseases or colon cancer. Their primary food was buffalo meat with the fat, both of which have zero fiber. They also ate pemmican, a mixture of dried buffalo meat crushed and shredded before being mixed 50/50 with melted buffalo fat. The fat content was about 70% on a calorie basis. Dried berries were sometimes added, but this was a small part of the caloric intake. The pemmican would keep for years without refrigeration and provided complete nourishment. The Hudson Bay Company and North West Company (fur trappers and traders) purchased pemmican from the Indians by the ton and even had a specification whereby they would pay a higher price for premium pemmican made with bone marrow fat. The Indians and European explorers to North America also preferred the buffalo tongue for consumption after a kill due to its high fat content.

Bad bacteria and yeast in the colon turn dietary fiber into noxious chemicals.

Vegetarians promote the fiber theory because meat does not contain fiber. But why do medical professionals and pharmaceutical companies promote the fiber theory in light of

Chapter 2

overwhelming scientific evidence against dietary fiber? Why do gastroenterologists promote the high-fiber, low-fat diet that does not relieve patients of their dreadful symptoms? Why must patients remain under their care for 30 or 40 years with digestive problems, constipation, diarrhea, bleeding ulcers, diverticulosis, fissures, and an annual colonoscopy? The harsh immunosuppressant drugs control the autoimmune diseases somewhat, but the patients suffer year after year as the diseases get worse and worse. The side effects from the drugs cause degenerative disc disease and osteoporosis, and allow other diseases to flourish because the immune system is suppressed. Could it be because their patients remain under their care for 30 or 40 years? Eventually the patient dies in old age from colon cancer. Many also suffer the horrible experience of colon removal and the installation of an external fecal collection bag. The medical community considers this treatment to be successful. On the following pages, you will read testimonies from people with ulcerative colitis and Crohn's disease who had significant improvements in their digestive health after switching to an ultra low-fiber, high-fat diet that is directly opposite to the one recommended by their gastroenterologists.

Proof Dietary Fiber Is Unhealthy

Low-Carbohydrate Diets Incorrectly Promote Dietary Fiber

Robert C. Atkins, MD, cardiologist and nutritionist (1930 – 2003), authored the very popular low-carbohydrate book, *Dr. Atkins' Diet Revolution* and several others. He came under constant attack with criticisms that his diet was deficient in fiber, but he countered these attacks with the claim that the low-carbohydrate vegetables contain large amounts of fiber. The attacks were driven by the fact that he recommended eating meat which contains no fiber. Dr. Atkins specialized in heart disease, diabetes, and cancer—not inflammatory bowel diseases and autoimmune diseases. Just prior to his death, he became suspicious that there was a negative connection between dietary fiber and intestinal diseases, and he made comments on his website that were mildly critical of dietary fiber. It is unfortunate that his website was shut down after his death. He never reached the point of fully identifying dietary fiber as a serious health hazard.

Other nutritionists and doctors who promote a low-carbohydrate lifestyle either promote dietary fiber as well or simply ignore the topic. They simply do not recognize the negative connection between dietary fiber and autoimmune diseases.

Dietary Fiber Does Not Relieve Constipation.

Medical schools teach that fiber in the colon absorbs water to form a soft stool that relieves constipation. This part of the fiber theory is repeated over and over by medical schools, fiber supplement salesmen, and gastroenterologists. The theory is so widespread and widely accepted by the medical community that they never bother to test it. The manufacturers of fiber supplements are certainly not going to test the fiber theory.

Chapter 2

There is one big problem with the theory that a high-fiber diet relieves or cures constipation. Most people who suffer with chronic constipation say the extra fiber doesn't help. In fact, most people on Internet message boards and forums claim the high-fiber diet and fiber supplements make their constipation worse. The manufacturers of fiber supplements reluctantly admit this happens. Patients of dieticians and gastroenterologists often admit that the high-fiber foods increase constipation or fail to help in any way. The following statements are typically made by people who promote the high-fiber diet to relieve constipation or diarrhea, but they often warn that the high-fiber foods can make these conditions worse. This is strange logic.

- Some people don't tolerate fibrous foods well.
- Fiber supplements in the diet can cause intestinal gas (flatulence), diarrhea, constipation, bloating, and other digestive discomforts.
- Take fiber supplements or laxatives when your daily consumption of high-fiber foods does not relieve constipation.
- Take fiber supplements such as psyllium mucilloid, polycarboxisal, or hemicellulose to relieve diarrhea. They absorb water and produce bulk in the stool.
- Drink 10 glasses of water each day when using fiber supplements because the fiber may cause constipation to be more severe.

The fiber theory is wrong. Constipation is not caused by the lack of fiber in the diet. Constipation is caused by the body drawing water out of the colon. Don't ask your gastroenterologist about this. He was never taught it in medical school.

The colon draws water out of the stool in an attempt to stop fermentation of the fiber and carbohydrates by the pathogenic

Proof Dietary Fiber Is Unhealthy

bacteria and yeasts. Eating a high-fiber diet in conjunction with fiber supplements does not work because it simply promotes continual fermentation. The colon becomes a hazardous waste dump. Constipation tends to slow down the reaction. It is no surprise that some people don't tolerate fibrous foods well. That should be expected, since high-fiber foods are always high-carbohydrate foods.

The Dietary Fiber Theory is pure conjecture. It is not scientifically true.

Medical schools and the professional medical societies in the United States are too large, interdependent, and cumbersome to ever change. A large study on dietary fiber would simply be ignored as other studies have been. For this reason, patients are turning to alternative medicine in droves.

Dietary Fiber Does Not Relieve Diarrhea

Many people tolerate constipation for years or even decades. They simply take laxatives, drink prune juice, take laxative herbs, fiber supplements, or use other temporary measures to gain relief, but the game is over when the constipation turns to chronic diarrhea. Some people are forced to use the bathroom 20 times a day and numerous times during the night. The urge to go never subsides between emergency dashes to the toilet.

It may be possible for those with chronic constipation to go shopping, go to work, and lead a fairly normal life. The same is not true with chronic diarrhea. People on Internet message boards and forums frequently report the highly embarrassing accident of messing their pants in the car on the way to work or halfway in their dash to the bathroom. Adult-size diapers are their only solution.

Chapter 2

Medical schools teach that fiber in the colon absorbs water to form a soft stool that relieves diarrhea. They claim that fiber is the miracle cure-all for diarrhea, constipation, and other gastrointestinal problems. This part of the fiber theory is repeated over and over by medical schools, fiber supplement salesmen, and gastroenterologists. The assertion that dietary fiber cures constipation and diarrhea is false.

When drawing water out of the colon doesn't stop the fermentation of fiber and carbohydrates by pathogenic bacteria and yeasts, the body takes the opposite approach. FLUSH time! The body simply floods the colon with water in order to rid itself of the problem, and it doesn't care whether you are stuck in traffic, attending a board meeting, or speaking before a group. When the colon flush comes, you go. Taking more fiber supplements does not work to stop chronic diarrhea, and eating high-fiber, high-carbohydrate foods doesn't work either.

My diet program provides perfect nutrition and deprives the pathogenic bacteria and yeasts of their favorite foods, fiber and carbohydrates. My diet program eliminates those foods that trigger an autoimmune disease flare; consequently, the colon settles down without constipation or diarrhea. The offensive flatulence is eliminated because digestive gasses are produced by the bacteria and yeasts. You can read the entire diet requirements in detail in the Chapter 5 titled, "Achieving Remission in Asthma, Inflammatory Bowel Diseases, and Other Autoimmune Diseases."

Proof Dietary Fiber Is Unhealthy

Perfect Poop

The ultra low-fiber, low-carbohydrate, high-fat, high-protein diet suppresses the pathogenic bacteria and yeasts in the colon to produce perfect bowel movements. The stool is firm but soft and passes easily. The urge to go is slight, and waiting for a convenient time is very easy. The claim that a person should have one or more bowel movements each day is a myth. Skipping a day is normal and does not cause a plugged feeling. Straining is not required to have a normal bowel movement.

The stool is a perfect gauge of digestive health. Perfect poop produced by proper digestion will be medium brown in color and sink in the bowl. The stool will retain its shape and not disintegrate into fragments. The volume should be considerable and of medium diameter. "Pencil poop" indicates the colon is constricted from inflammation. Perfect poop will leave relatively clear, not cloudy, water in the toilet bowl. Since it will have very little or no odor, there will be no offensive smell in the bathroom. Flatulence during the day will be infrequent with very little or no odor. Fiber and carbohydrates in the diet cause excessive, smelly flatulence because the bacterial fermentation in the colon and the lower part of the small intestines produces gas. A floating stool indicates the improper digestion of fats, but this is not an unhealthy condition. Undigested fats in the stool do not cause any problems whatsoever. Taking a lipase enzyme may help, but pancreatic enzymes can result in more gas being entrained in the stool. The gas will cause the stool to float, but this is not of any concerns.

Red color in the stool indicates blood is being released near the end of the colon or rectum. Black color indicates a blood release farther up in the digestive tract or stomach. Yellow could be undigested fat. Green indicates excess bile acids or a short travel

Chapter 2

time. The normal color is brown to dark brown as everyone knows. The diet program in the IBD chapter will produce proper digestion and relieve constipation and diarrhea.

Fecal Transit Time Nonsense

Most professional gastroenterologists and other proponents of the high-fiber diet claim that fecal matter is a highly toxic waste that will cause cancer, diverticulitis, and ulcerative colitis if the transit time in the colon is not rapid. They refer to the fecal matter as a hazardous waste that will kill you if you don't have several bowel movements each day. They insist that you eat a high-fiber diet in order to increase the bulk of the stool and the fecal transit time. This theory is pure nonsense. In fact, it is backward.

Fecal matter is always in the colon and in contact with the lining. A fast transit time does nothing to reduce this contact as many people falsely believe. For example, your hand will hurt if you slowly pour hot water over it. Pouring the water fast will not make the hot water any less painful. In fact, the fast-flowing water will hurt worse. The same is true of the colon. If fecal matter were truly toxic, the fast movement would not make it any less toxic and would not reduce the exposure. The contact is always present, 100% of the time.

Eating high-fiber foods and taking fiber supplements are very hazardous to your digestive health. Taking laxatives to increase the fecal transit time is even worse. The current epidemic of inflammatory bowel diseases is directly related to the increase in dietary carbohydrates, fiber, and fiber supplements. The high-fiber theory is being promoted everywhere you turn, and it is slowly killing people.

Proof Dietary Fiber Is Unhealthy

A low transit time is healthier. A low transit time allows all of the food to be digested in the small intestine. Undigested food in the colon results in a proliferation of pathogenic bacteria—the true cause of diverticulitis and ulcerative colitis. A fast transit time causes many diseases because of the overgrowth of pathogenic bacteria, viruses, yeast, fungus, protozoa, and parasites in the colon. A fast transit time turns the colon into a toxic fermentation tank. This is not normal or healthy for humans. Cows have a digestive system based on fermentation. Humans do not.

The low-carbohydrate, low-fiber diet presented here promotes optimum digestion with a slow transit time, very little flatulence, and a soft, firm stool. Skipping several days without a bowel movement is perfectly normal and healthy.

Dietary Fiber Causes Bowel Diseases and Cancer

Glucose molecules attach together in long chains called *carbohydrates*. The body breaks down these chains in the small intestine during digestion to allow absorption of the glucose into the blood. Dietary fiber consists of tightly-bound carbohydrate chains. The body cannot break down these large fiber molecules during digestion because we do not have the necessary enzymes. Fiber typically passes through the small intestines intact. Most doctors and nutritionists believe the fiber continues unbroken through the colon and passes out of the body in the stool. This assumption is false. The bacteria and yeasts in the colon easily break down the fiber which then becomes the major food source for these pathogenic bacteria and yeasts. Healthy individuals may not be able to digest dietary fiber, but pathogenic intestinal bacteria and yeasts certainly can. These bacteria and yeasts ferment the fiber to produce intestinal gas, alcohol, acetaldehyde, lactic acid, acetic acid, and a host of other toxic chemicals that damage the lining of the colon. The toxic chemicals cause leaky

Chapter 2

gut syndrome, inflammatory bowel diseases, and a host of other autoimmune diseases. Yeasts also feed on the fiber, and the result is a systemic yeast infection.

Most of the digestion and absorption of nutrients occurs in the small intestine. Its long length alone provides a large surface area for digestion and absorption, and that area is further increased by mucosa villi on the surface of the interior intestinal wall. The mucosa (inside surface) and the submucosa allow the small intestine to complete the process. The mucosa forms a series of villi (small projections) only 0.5 to 1.0 mm high, which give the intestinal mucosa a velvety appearance. The villi are the primary structures responsible for the digestion and absorption of food. These villi become damaged from a diet of excessive carbohydrates and the deficiency in proteins and fats. A protein deficiency may be the key because the villi have a short life and are continually being replaced. This replacement may be prevented when one or more of the essential amino acids are missing from the diet. Digestion of the carbohydrates in the small intestine can then become incomplete, and undigested carbohydrates pass to the colon where they do not belong. This presence of carbohydrates in the colon encourages the growth of pathogenic bacteria (disease-causing bacteria such as Clostridium difficile) and pathogenic fungi, which are opportunistic. The healthy intestinal bacteria can become overpowered and destroyed by the bad bacteria and/or candida yeast infections. Treatment of the bacteria with drugs is only a temporary solution because the opportunity exists for their re-establishment. Proper diet combined with the re-establishment of a good bacterial flora is essential. The bacteria must be starved to death by removing their primary food source—carbohydrates and fiber—from the diet. Removal of these undesirable bacteria by drug treatment will not be successful if carbohydrates

Proof Dietary Fiber Is Unhealthy

continue to be consumed. They will simply return after drug treatment ends because they are opportunistic.

When many IBD sufferers test their fecal matter, the results indicate they have one or more pathogenic bacteria in the colon. They then make the assumption that they must have contracted the pathogenic bacteria in some manner and assume that this is the cause of their inflammatory bowel diseases. Both of these assumptions are wrong. People are routinely exposed to these bacteria, which are suppressed by a healthy digestive system. The abundance of carbohydrates and fiber in the diet permits the opportunistic pathogenic bacteria to proliferate and make their IBD worse. Pathogenic bacteria proliferate in the colon and appendix, where they set off chronic inflammation due to the consumption of high-carbohydrate foods, such as whole grains, sugars, and fruit. Appendectomies are very common in people with IBD.

> **Fiber and carbohydrates turn the colon into a fermenting hazardous waste dump.**

Carbohydrates trigger an increase in insulin, which promotes the release of epinephrine (adrenalin), and the result is a collection of physiological actions called the *fight-or-flight* responses. Two of these responses are in the gastrointestinal tract where muscular movements and digestive secretions slow down or even stop. This produces constipation. Most gastroenterologists erroneously recommend a high-fiber diet to combat the constipation brought about by carbohydrates, when the correct action should be to stop eating carbohydrates. The fiber can be digested by pathogenic bacteria, and it encourages their growth. Diarrhea results because these bacteria stimulate intestinal secretions, and this is why so many people with intestinal

Chapter 2

diseases find no relief from eating fiber or taking fiber supplements. High-fiber foods are always high-carbohydrate foods. The body responds to prevent offensive carbohydrates from entering the colon by pulling out the water. This causes constipation. The body can also reverse this tactic by pouring lots of water into the colon as it tries to flush out the offensive carbohydrates, bacteria, yeasts, and their toxic byproducts. This causes diarrhea, and the vicious cycle is underway.

The low-carbohydrate, low-fiber diet recommended here has been shown to stop this vicious cycle. My high-protein, high-fat diet reduces insulin and adrenalin responses and promotes healing, healthy digestion, and normal bowel movements. In most cases, these improvements occur in a matter of weeks or days simply by eating a diet typical of many primitive cultures throughout the centuries. This was the natural diet for mankind prior to the cultivation of grains and the production of factory foods. In order to fully appreciate this diet plan, you must overcome your belief in the current nutritional myths promoted by food manufacturing companies.

Some people have a defective gene that causes carbohydrate intolerance—the inability to absorb glucose and galactose in the small intestine. The unabsorbed glucose and galactose continue through the intestinal tract to the colon where they promote the growth of pathogenic bacteria, fungi, and yeasts. Glucose-Galactose Malabsorption is inherited as a recessive genetic trait. Sodium-glucose cotransporter [SGLT1] is the gene responsible for this defective disorder. It is located on the long arm of chromosome 22 (22q13.1).

Dietary fiber does not prevent colon cancer as commonly claimed. Scientific studies have proved that fiber increases colon cancer. The abundance of carbohydrates and fiber in the diet

Proof Dietary Fiber Is Unhealthy

allows the opportunistic pathogenic bacteria to proliferate and thereby make the inflammatory bowel diseases worse.

Dietary Fiber Depletes Vitamin and Minerals from the Body

Fiber absorbs vitamins and minerals and discharges them from the body. Calcium and magnesium are required to prevent osteoporosis, bone loss, hip fractures, and degenerative disc disease; fiber leaches them from the digestive tract and discharges them in the stool. Fiber increases the risk of osteoporosis.

Do not take fiber supplements. Do not take psyllium seed husk which is very abrasive to the digestive system. Do not eat wheat bran or rice bran. Fiber is a bad dude. Avoid all whole grains, brown rice, fruit, and dried beans as they are high in both fiber and complex carbohydrates—a double blow to the digestive system.

Dietary Fiber Supporters Have an Agenda

Vegetarians love Dr. Burkitt's fiber theory because meat has no fiber. They use the fact that meat contains no fiber to blame meat for every ailment known to mankind. Every vegetarian book and website is plastered with this false claim. It is ironic that about 80% of the people on inflammatory bowel disease Internet message boards and forums are vegetarians or previously ate a highly vegetarian diet, yet vegetarians are only 6% of the population in the United States. Bowel disease and colon cancer have become epidemic in vegetarians in all English-speaking countries that follow the low-fat, high-fiber diet recommendations.

Chapter 2

I know only two vegetarians by name. Scott was a 50-year old, healthy-looking landscaper. He was trim, and his weight was perfect for his size. When we discussed health and diet in 2002, he rejected my low-carbohydrate diet comments and stated he was a vegetarian. He had just formed a vegetarian support group, but his diet did not treat him well. Scott was diagnosed with inoperable colon cancer in 2005. Scott died in June 2006.

Exercise will most likely increase the risk of cancer rather than eliminate it. Avoiding red meat and animal fats will not eliminate the risk of cancer. A vegetarian marathon competitor sadly discovered this fact. Nancy participated in marathon events even though she was 49 years of age. She developed a persistent cough that her doctor said was due to post nasal drip. When the cough became worse, she visited another doctor; her chest x-ray revealed the dreaded news. Nancy had tumors in her lungs that proved to be cancerous. Chemotherapy seemed to place the cancer well into remission. The tumors shrank for a while but recurred. Nancy was given every new experimental cancer drug available; nothing worked and the side effects were horrible. She had chemo a second time, although doctors rarely try it. It did not help her. Nancy didn't realize that excessive exercise literally wipes out the immune system. It is common knowledge that marathon runners easily get sick after a race because their immune systems are shot. A strong immune system is necessary to prevent cancer. She was a vegetarian and didn't realize that all immune system cells are made from amino acids such as those found in meat. Nancy had an unprecedented third round of chemotherapy. Nancy died in December 2006, approximately two years after her original cancer diagnosis.

Dietary supplements will not kill cancer cells. Wild claims made by those who sell herbal products are simply not true. Herbs will

Proof Dietary Fiber Is Unhealthy

not kill cancer cells. Chinese medicine will not kill cancer cells. Homeopathic products will not kill cancer cells. Acupuncture will not kill cancer cells. Magnetic therapy will not kill cancer cells. All of these treatments are frauds.

The modern supermarket displays thousands of high-carbohydrate, high-fiber foods which occupy 80% of the floor space in the store. These foods include whole grain breads, bagels, cereals, pasta, rice, legumes, potatoes, chips, cookies, candy, fruit, and fruit juices. The promoters of carbohydrates and fiber have caused the current epidemic of constipation, diarrhea, inflammatory bowel diseases, diabetes, heart disease, and cancer.

Whole grain cereal manufacturers are major supporters of the false fiber theory because cereal grains are high-fiber foods. They are constantly spouting claims of fiber health benefits without providing any supporting science. The supporters of the fiber theory simply reference each other and Dr. Burkitt's erroneous study. The new studies supporting dietary fiber cannot be trusted.

Humans Are Not Vegetarians like Cows, Goats, and Deer

Vegetarian animals such as cows, goats, and deer digest food by a fermentation process. Bacteria in the stomachs of these animals are able to break down the fiber and cellulose to form glucose for their food. The bodies of the dead bacteria are composed of protein and fat which then become available to the cow or goat. Goats love to eat paper because paper is an excellent food for a goat. If you have visited an animal petting farm, you can testify that a goat snatched the information sheet out of your hand and ate it. Paper is cellulose that is digested in the fermentation

Chapter 2

process within the goats' stomachs. Humans don't eat paper because we do not have a vegetarian digestive system.

Cows and goats produce large quantities of milk consisting of generous amounts of protein and fat, yet the grass they eat has very little protein and fat. Grass is primarily complex carbohydrates in the form of fiber and cellulose. Deer love to eat small twigs, buds, and tree leaves for the same reason. Protein and fat are not contained in their food, but the fermentation process in the animals provides all the protein and fat they require.

Humans do not have a vegetarian digestive system. Humans cannot digest fiber or cellulose. However, the bad bacteria in the human colon can certainly feed on fiber and cellulose, which leads to inflammatory bowel diseases. Paper is wonderful food for goats but not for people, who have carnivorous digestive systems almost identical to those found in cats, dogs, wolves, tigers, lions, etc. Protein and fat are the best human foods because they are essential macronutrients. Carbohydrates are not classified as essential macronutrients.

Unhealthy High-Fiber Foods

Bran Cereal	Oats	Rice	Soybeans
Legumes	Split Peas	Lentils	Apples
Broccoli	Potatoes	Grapefruit	Pears

Absolute Truth Exposed

Volume 1

Chapter 3

Top Ten Nutritional Myths, Distortions, and Lies That Will Destroy Your Health

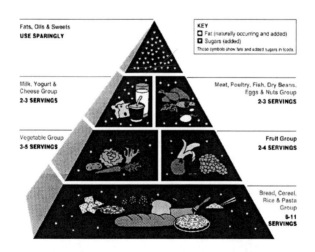

USDA 2004 Food Guide Pyramid
Carbohydrates = 60%; Fats = 30%; Protein = 10%
Warning! Eating this diet for 20 years will cause many diseases.

Chapter 3

This book will prove that eating red meat and natural animal fats while restricting carbohydrates is not only healthy but will prevent and cure many diseases.

The professional dietary and medical advice given by the United States Department of Agriculture (USDA) and the United States Food and Drug Administration (USFDA) have been an utter failure. The USDA has published the official diet in the form of an icon called the *USDA Food Guide Pyramid*. The USFDA has published the official *USFDA Recommended Daily Allowance* (RDA) for carbohydrates, proteins, and fats. However, the recommendations have a big flaw. Neither the *Food Guide Pyramid* nor the RDA is based on scientific facts. If you follow the *Food Guide Pyramid* as your daily nutritional guide, you will encounter many related degenerative diseases, including diabetes, heart disease, cancer, bowel disease, autoimmune diseases, and many more.

Are the foods we are told to eat making us sick? Yes, absolutely!

The *USDA Food Guide Pyramid* and *USFDA Recommended Daily Allowance* are solely responsible for the current high incidence of adult onset Type II diabetes, which has tripled in the last 30 years. Type II diabetes is becoming common among teenagers, who were once thought to be excluded from this "age related" disease. People with diabetes have a significantly increased risk of cancer. Now the USDA has the gall to tell us it was our fault for not exercising enough. They were wrong then, and they are wrong now. Exercise will not prevent the escalating health-care crisis.

Top Ten Nutritional Myths

Healthy High-Protein and High-Fat Foods

All Fresh Meat	Beef	Lamb
Pork	Chicken	Turkey
Wild Game	Fish	Seafood
Shell Fish	Hard Cheeses	Coconut Oil
Butter	Organ Meats	Eggs

USDA 2005 Food Guide Pyramid
Source: United States Department of Agriculture
Carbohydrates: 50% to 60% depending on the individual

The new *USDA 2005 Food Guide Pyramid* adds more confusion because it contains many different pyramids from which to select. This system will cause the current obesity, cancer, diabetes, and heart disease epidemics—that are a result of the previous *Food Guide Pyramid*—to become more severe.

The USFDA nutritional guidelines are not scientific.

Many people think the Atkins' low-carbohydrate diet is lacking essential nutrients because it doesn't match the results of the *Food Guide Pyramid*. They reference the US Food & Drug Administration's (USFDA) *Nutritional Guide for Daily Values* (DV)

Chapter 3

as shown on all nutritional labels. The US Department of Agriculture *Food Guide Pyramid* was developed by vegetarians with an agenda. Nathan Pritikin and Senator George McGovern were a major influence—if not the perpetrators. More will be said about Nathan Pritikin and Senator George McGovern in Chapter 4.

There is no science behind the *Food Guide Pyramid*. It was a scam from the beginning—a make-believe nutritional plan to limit the consumption of animal products. The results have been rampant heart disease, cancer, diabetes, osteoporosis, intestinal diseases, and a medical handbook full of other ailments. The *USFDA Nutritional Guide* is based on the *Food Guide Pyramid*. This is easy to verify. Simply go to a food count book and enter a 2000-calorie diet exactly according to the pyramid. The results will show every nutritional requirement to be perfectly achieved. It's all a scam. The daily values for vitamins and minerals were simply set to match the food in the pyramid. There is no hard science behind the establishment of the USFDA daily nutritional requirements. Eicosapentaenoic acid (EPA) and docosahexaenoic fatty acid (DHA) omega-3 fatty acids—commonly called *fish oils*—are essential nutrients in the human diet. You will develop diseases and possibly die without them. Yet, they are not even listed on the US Food & Drug Administration's *Nutritional Guide for Daily Values*. Pick up any food package and look at the nutritional label. They are not there. The amount of saturated fat is given, but saturated fat is not an essential nutrient. There is no minimum requirement. A maximum requirement of 10% is listed, but that value is wrong. It is not based on proven science.

Professional nutritionists argue to the death in support of the minimum daily nutritional values but will supply no solid science to support them. Their arguments are based on theories, conjecture, or outright lies. Unbiased research, the study of

Top Ten Nutritional Myths

primitive societies, and eating a low-carbohydrate diet easily prove the minimum daily nutritional values are fraudulent. For example, nutritionists say that fresh meat contains no vitamin C, and medical societies claim that a vitamin C deficiency causes scurvy. Books that list the amount of vitamins and minerals in typical foods do not list vitamin C in meat. Professional nutritionists and USDA health publications claim that fruits and vegetables must be eaten to prevent scurvy. It is easy to prove that these claims are false. The Eskimos lived for centuries on an all-meat and animal fat diet of seal, polar bear, caribou, fox, wolf, whale, fish, and fowl without developing any signs of scurvy. White men lived with them for many years in the early 1900s without showing signs of scurvy, thus proving the absence of scurvy was not racial. In fact, sailors on whaling vessels often cured their scurvy by eating fresh meat. Fresh meat is the only food that provides complete nutrition. The concept that a healthy diet requires eating from four or five food groups daily is pure nonsense without any scientific basis. Nutritional and medical societies simply ignore the scientific facts.

Vitamin and mineral food count books cannot be trusted to give the correct amounts of vitamins and minerals contained in various foods. Meat is typically ignored as a recommended vitamin or mineral source in favor of fruits and vegetables. In truth, red meat, fowl, fish, and other seafood are packed with vitamins, minerals, essential fatty acids, and essential amino acids. Meat contains every requirement needed in the human diet. Meat is the perfect food.

Vitamin B12 is only found in meat. When explorers discovered the North American Plains Indians and Eskimos, they found them to be in excellent health even though their diet consisted almost entirely of meat. Nutritionists of that day and modern nutritionists cannot concede the fact that these people did not

Chapter 3

suffer from scurvy because of Vitamin C deficiency. The explorers found that fresh meat and pemmican not only prevented scurvy but cured it in those already suffering from the disease. The Indians and Eskimos lived in excellent health on an all-meat diet. The Eskimos who lived in Northern Canada on the Mackenzie River Delta ate a diet consisting almost entirely of salmon, caribou, and seal.

The North American Plains Indians lived primarily on pemmican, a mixture of lean buffalo meat that was dried, shredded, or pounded fine and mixed with melted fat. The fat was half internal hard fat and half external soft back fat. This diet produced tough, strong, and perfectly healthy children. The content was 80% fat on a calorie basis, contrary to the erroneous statements made in numerous books that the Indians ate lean meat. These writers came to this hasty conclusion because buffalo muscle is leaner than beef. The Indians added 50% fat on a volume basis to make the pemmican. The fat was found in thick slabs on the animals' backs between the hide and the flesh.

> **We are approaching the 60-year anniversary of the low-fat diet craze.**
> **Have you been to the beach lately?**

The Hudson Bay Company and North West Company purchased pemmican by the ton from the Indians to supply the fur trappers. The United States Government later moved the Indians to reservations and provided them with grain, flour, and sugar for food instead of meat. The high-carbohydrate diet supplied by the government destroyed the health of the Indians by creating an epidemic of obesity, diabetes, heart disease, and cancer that continues to this day.

Top Ten Nutritional Myths

Beginning in 1906, Arctic explorer Vilhjalmur Stefansson and his companion, Karsen Anderson, spent several years living with the Eskimos in Northern Canada. They lived on an all-meat and fat diet of seal, polar bear, caribou, and fish, which doctors and nutritionists insisted was impossible. Stefansson could observe no adverse health effects from the lack of carbohydrates and fiber in their diet. In fact, their health was excellent with absolutely no dental caries, no heart disease, no cancer, and excellent bone health. To prove his statements were true, both men entered Bellevue Hospital in New York City to live for a full year on a supervised diet of meat and fat only. Stefansson was allowed to leave the hospital to attend to personal matters with the promise to continue the strict diet. He did so. At the end of the year, both men had improvements in health as measured by all the tests available at the time. Stefansson wrote about his experience and the test results in a three part report that was published in *Harper's Monthly Magazine*, November and December 1935 and January 1936.

For centuries, the Eskimos lived on the Arctic ice north of Canada in the winter—not summer—spearing seals through their breathing holes in the ice. The diet of the people and dogs was 100% seal meat and seal fat consisting of approximately 80% fat and 20% protein. Their carbohydrate intake was zero—none. They did not have problems maintaining blood sugar levels. Dietary carbohydrate is not needed for the brain, heart, or any other organs because the body can make glucose from fat and protein. The blood sugar level is maintained perfectly well—naturally—on the low-carbohydrate diet.

When Eskimo women gave birth, their breast milk had the same composition as that of any other mother: 58% fat, 12% protein, and 30% carbohydrates (lactose or milk sugar). The weaned infants ate the adult Eskimo all-meat diet which contained 80%

Chapter 3

fat, 20% protein, and zero carbohydrates. The children grew to be muscular, strong, and perfectly healthy. The Arctic Eskimos lived in excellent health on a diet of 100% fresh seal meat and fat. You may want to read Mr. Stefansson's book, *MY LIFE WITH THE ESKIMO*, if you can find a used copy.

Captain Robert Falcon Scott, Royal Navy (June 6, 1868 - March 29, 1912), was a British Naval officer and Antarctic explorer. Scott and his party narrowly failed to be the first to reach the South Pole, having been preceded by the Roald Amundsen team. The Scott party all died on the Ross Ice Shelf while attempting to return to the safety of their base. Scott has become the most famous hero of the "heroic age" of Antarctic exploration. The Scott party died because they carried the wrong food with them. They brought cooked and salted lean meat that was prepared in England. The meat was low in fat and lacked the calories needed for such an expedition. The natural vitamin C in fresh meat was destroyed in the processing, and the explorers suffered severely from scurvy. They would have been well nourished if they had prepared dried, fresh, lean meat with an equal amount of rendered fat similar to pemmican made by the North American Plains Indians. They would have also been well nourished had they packed the sleds with fresh penguin or seal meat. The penguins were practically walking around between their legs when they landed at the Antarctic. The explorers were literally stumbling over excellent food and didn't know it.

The ice along Northern Canada and Alaska has always melted at the shoreline in summer. That is why explorers tried to search for the Northwest Passage—an assumed ice free route around Northern Canada. The passage is rarely fully open, but an icebreaker has made the trip from Greenland to Alaska. Modern small boats can often succeed in the passage with good luck and excellent weather. The Franklin party, comprised of two ships,

Top Ten Nutritional Myths

tried to find the passage in 1825. Because they knew they would be trapped in the ice all winter, they brought guns, food, and lime juice for vitamin C. The Royal Navy and the Merchant Navy supplied all the British sailors with lime juice to prevent scurvy. They all died of scurvy anyway because vitamin C decomposes over time when exposed to air, and the copper tubing through which the lime juice was piped destroyed the vitamin C.

The men on the trapped ships could see the nearby Eskimos waving as they went about their daily routine hunting seals, raising babies, taking care of old people, and feeding their dogs. The Eskimos had no guns and ate a 100%-meat diet. Professional nutritionists have brainwashed the public by claiming that meat has no vitamin C. Actually, the Eskimos' 100%-meat diet prevented scurvy and cured those who suffered from it as Arctic explorer Vilhjalmur Stefansson proved. Read his book, *Great Adventures and Explorations,* if you can find a used copy.

Many large food manufacturing organizations have lobbied the US Congress and put money into newspapers, books, universities, nutritional organizations, and medical organizations during the last 60 years in order to convince the consumer these myths are true. You can test yourself to see the degree of brainwashing that has crept into your thinking. Extensive study beginning with the information presented in this book is required to get past the false information coming from these organizations, politicians, and religious vegetarians. Reading official government publications on diet and health is the worst approach to learning the truth.

Chapter 3

This book will prove the healthiest diet for humans is:

70% total fat on a calorie basis
    31% saturated fat
    7% polyunsaturated fat
    25% monounsaturated fat
    7% other fats
27% protein
3% carbohydrates (20 gm, of which 3 gm or less is fiber).

**Boiled Salmon and Asparagus with
Real Butter and Potassium Chloride Salt
Supplemental Vitamins Are Taken to
Provide Overwhelming Nutrition**

Top Ten Nutritional Myths

Myth No. 1

Saturated Animal Fats Cause Heart Disease and Cancer

This **myth** is so widespread and often repeated that many people believe it without question. However, a quick review of the health and diets of people in 1900 reveals they ate a very high level of saturated fat from meat, butter, and lard but had virtually no heart disease or cancer. Proponents will cite a study showing that an increase in saturated fat in the diet raises the blood cholesterol level and claim that elevated cholesterol is a precursor of heart disease. Other studies claim saturated fats increase the risk of breast cancer in premenopausal women. These studies never blame the high levels of carbohydrates which are always included. They cite studies indicating that some country consumes a higher amount of saturated fat than another country and has a greater incidence of some form of cancer while they ignore vast differences between the two countries. These are big fat lies backed by slick schemes to distort the truth.

> **Cholesterol in eggs and saturated fats in red meat do not cause heart disease.**
> **Carbohydrates and insulin cause heart disease.**

One study is nicknamed the *French paradox* because the people of France ate a diet very high in saturated fat but had a heart disease rate only one third that of the United States. The French people have a much lower intake of carbohydrates and a higher intake of saturated fats—the real reason for their superior health. The French are noted for the high amount of cheese in their diet. Cheese has no carbohydrates but is high in saturated fats. Cheese is a perfectly healthy food as the French have proved.

Chapter 3

The National Heart, Lung, and Blood Institute (NHLBI) Framingham Study has been ongoing since 1948. According to NHLBI Director, Dr. Claude Lenfant, "This study suggests that obesity is an important risk factor for heart failure in both women and men." The study found a small correlation between heart disease with elevated LDL cholesterol and total cholesterol; however, these are dependent variables because high cholesterol is associated with obesity. The Framingham study also found that those who ate the most saturated fat, the most calories, and the most cholesterol were the most physically active. They also weighed the least and had the lowest levels of serum cholesterol. The people who ate the most saturated fat were the healthiest and had the lowest risk of heart disease.

"NHLBI Framington Heart Study"
http://www.framinghamheartstudy.org/

"NHLBI's Framingham Heart Study Finds Strong Link Between Overweight/Obesity And Risk For Heart Failure."
http://www.nih.gov/news/pr/jul2002/nhlbi-31.htm

People Lie About Saturated Fat for Two Major Reasons.

- Vegetarians don't want other people to kill animals and eat the fatty meat.
- The edible oil industry wants people to buy more of their vegetable oil products.

Vegetarianism is a religion falsely disguised as a healthy way of eating. True vegetarianism is animal worship. These people place animals on the same value level as human beings and sometimes above that of the human race. The false information they spread about healthy eating has led to a dramatic surge in obesity, diabetes, inflammatory bowel disease, heart disease and cancer—not found in low-carbohydrate, high-fat, meat-eating

Top Ten Nutritional Myths

societies. The low-fat, high-carbohydrate diet craze of the last 60 years has resulted in untold suffering and the early deaths of millions of people, both of which continue unabated to this day.

The edible oil industry produces omega-6 polyunsaturated vegetable, seed, and grain oils such as corn, soybean, safflower, sunflower, peanut, and cottonseed. These oils are thought to be among the leading causes of heart disease and cancer, both of which increase in concert with increases in omega-6 fatty acids in the diet. Omega-6 fatty acids are pro-inflammatory and have been proven to cause or contribute to a long list of autoimmune diseases, such as rheumatoid arthritis, lupus, multiple sclerosis, Crohn's disease, fibromyalgia, irritable bowel syndrome, inflammatory bowel disease, psoriasis, and many others. Check out the follow study that associates vegetable oils with postmenopausal breast cancer.

> "Results: Saturated fat and the omega3-omega6 fatty acid ratio were not related to increased risks, but positive trends were seen for total ($p = 0.031$), monounsaturated ($p = 0.002$), and polyunsaturated fat ($p = 0.0009$), especially omega6 fatty acids and the polyunsaturated-saturated fat ratio ($p = 0.004$). With mutual adjustment for different types of fat, an elevated risk remained significant in the highest omega6 fatty acid quintile (RR= 2.08, 95% CI 1.08-4.01)." Postmenopausal breast cancer is associated with high intakes of omega-6 fatty acids (Sweden). Malmö Diet and Cancer Study: Department of Medicine, Surgery and Orthopedics, Lund University, Sweden.

The above study proves that the saturated fat and omega-3 fat contained in red meat and fish were not associated with an elevated risk for breast cancer in postmenopausal women.

Chapter 3

However, monounsaturated omega-9 fat in olive oil, and polyunsaturated omega-6 fat in grains, seeds, and nuts increase the risk of breast cancer—especially omega-6 fat in grains, seeds, and nuts. Monounsaturated fatty acids in olive oil did not get a clean pass as falsely claimed by those promoting the *Mediterranean Diet.* These excellent results are opposite to the false claims promoted by vegetarians and manufacturers of high-carbohydrate foods. Beef, lamb, and pork with saturated fats are very healthy foods. Oils from whole grains, seeds, and nuts prove to be inflammatory and contribute to the increase in breast cancer.

Healthy Fatty Acid Foods

Animal Fats	Arachidonic Acid	Saturated Fats
Fish Oils	EPA Omega-3	DHA Omega-3
Butter	Coconut Oil	Monounsaturated

Unhealthy Fatty Acid Foods

Grain Oils	Seed Oils	Most Nut Oils
Vegetable Oils	Omega-6 Fats	Polyunsaturated
Trans Fatty Acids	Soybean Oil	Corn Oil

The edible oil industry also produces highly saturated (hydrogenated) forms of these oils. These hydrogenated oils are known by their opponents as the deadly *trans fats.*

Twisting a study to produce a lie is very easy. These fraudulent studies use a high-carbohydrate diet as the base and increase the saturated fat as the variable. The results will always show an increase in the triglycerides and LDL cholesterol. They blame the saturated fat when the real cause is the excessive amount of carbohydrates in the diet. Saturated fat in a low-carbohydrate

Top Ten Nutritional Myths

diet improves cholesterol ratios by lowering the triglycerides and raising the good HDL cholesterol.

Saturated fat in the diet is necessary for healthy body cells; it produces strong cell walls that would otherwise be soft and floppy. The saturated fat resists penetration of the cell by invaders, oxidation, and attack by free radical molecules, ions, and elements.

> **Study shows unsaturated fatty acids stimulate cancer cell growth while saturated fatty acids inhibit cancer.**

"Saturated fatty acid-induced apoptosis in MDA-MB-231 breast cancer cells. A role for cardiolipin."
http://www.ncbi.nlm.nih.gov/pubmed/12805375

> "Unsaturated free fatty acids (FFAs) stimulated the proliferation of human MDA-MB-231 breast cancer cells, whereas saturated FFAs inhibited it and caused apoptosis (process limiting growth of tumors)."

What? Did they say that saturated fatty acids in meat inhibited breast cancer cells? YES!

Government diet and health publications tell us to reduce our intake of saturated fats and increase the intake of polyunsaturated fats contained in whole grain, nut, and seed oils. This approach increases the risk of breast cancer as the above study shows. History has proved the connection as well. Breast cancer has become epidemic. Women worry and panic at the thought of seeing their annual mammogram report.

Chapter 3

Dr. Robert C. Atkins confirmed the same support for saturated fatty acids. The following quotations were taken from his now inactive website.

> "Saturated fat has gotten a bum rap. The real enemy of your health is trans fat, which lurks in many processed foods. Many people mistakenly think all fats are nutritional villains responsible for everything from coronary artery disease to obesity. To the contrary, many health problems may actually be caused by deficiencies of certain fats—even the much maligned saturated fats. Fats serve an array of important functions, including cushioning our organs against shock and insulating vital tissues against cold. They also influence body temperature regulation, pain sensitivity, appetite control and cognitive performance."

"Saturated fats, found in large proportions in meat, butter and other animal products, as well as in such tropical vegetable oils as palm and coconut, are usually solid at room temperature. Numerous studies show that these fats play many beneficial roles in our bodies, including enhancing the immune system, protecting the liver and helping build healthy bones."

The harsh criticisms of red meat and saturated fat are not supported by scientific research or anthropological surveys. Weston A. Price, DDS, traveled worldwide in the 1930s to investigate the health of primitive peoples who could not obtain foods from the western world. He and his wife, Florence, found that all of these primitive groups ate a diet very high in fat. Some primarily ate animal meat and fat while others primarily ate seafood. The two different diets provided equally good health. They were all extremely healthy, strong, robust, and had almost

Top Ten Nutritional Myths

no dental cavities. They all had broad dental arches (jaw shapes), and the women had very easy childbirths because of their broad pelvic structures. The children who moved to a modern society developed crowded teeth with many cavities, and the women suffered difficulties in childbirth similar to our present western society. The "civilized" Indians of northern Canada also suffered greatly from tuberculosis and were sent back to live with the primitives where the diet consisting of nearly 100% meat with the fat cured their disease.

Natural saturated fats don't cause heart disease or cancer and never did. It has all been a big fat lie.

A TV documentary was produced in the country of Georgia—formerly part of the Soviet Union—by a Professor from the University of Colorado in Boulder. Unfortunately, the date and reference to the documentary have been lost. The professor and his team visited the Georgian people living on the Caucasus mountain slopes at 8,000 feet elevation on small farm plots surrounded by high peaks. The terrain was similar to that of people living on small farms high in the Swiss Alps as reported by Dr. Weston A. Price, and their diet was identical to that of the Swiss. Reaching the area was very difficult even with the use of a four-wheel drive Jeep®. The amazing characteristic of these people was their longevity—living to 100 years of age was common. They raised farm animals and grew hay, vegetables, and some grain. Fruit was unavailable due to the high elevation, and their remote location made imports impossible. The ground floor of the house served as a barn to protect the animals from the harsh winters, and the people lived on the second and third floors. The commentator said they ate meat, butter, and high-fat yogurt with every meal. He described it as "very un-Boulder," referring to the liberal college community of Boulder, Colorado—

Chapter 3

where a low-fat, vegetarian diet philosophy is rampant. They made good use of their farm animal products by eating lots of meat, butter, and saturated fats. The remaining food consisted of a few seasonal vegetables and a small amount of bread made from coarsely-ground whole grain. They did not import high-carbohydrate foods, such as sugar, honey, white flour, orange juice, or fruit. They did not grow potatoes. They did not have omega-6 vegetable oils, trans-fats, or hydrogenated oil. They were lucky people because their food supply was the perfect example of the Atkins' low-carbohydrate diet. These mountain people of Georgia clearly demonstrate the long-term health benefits of the low-carbohydrate diet. A study of Georgian people found that those who ate the most meat and fat lived the longest.

Universities are centers of higher brainwashing.

French women have the lowest rate of heart disease in the Western world even though they consume high levels of butter, cheese, and animal fats. France is reported to have 265 brands of cheese that typically contain 65% to 70% fat on a calorie basis; 45% to 50% of these fats are saturated. They are healthier because of their high level of saturated fat and low level of sugar and refined carbohydrates in the diet. This high level of saturated fat and low heart disease rate has become known by the confused low-fat dietitians as the *French paradox*. Sadly, the French are turning away from their natural foods to manufactured high-carbohydrate foods.

The peoples of Thailand are another paradox. They have a very low level of heart disease and diabetes but consume exceedingly high levels of saturated fat in coconut oil and pork lard.

The Grecian *Mediterranean Diet* has been touted as healthy because it contains olive oil and fish. Indeed, fish is a healthy

Top Ten Nutritional Myths

food, but the Grecians consume high levels of saturated fat in feta cheese, butter, lard, and poultry fats. Those who claim the *Mediterranean Diet* is low in saturated fats are simply wrong. The Italians eat a similar diet, and both consume high levels of carbohydrates in the form of pasta.

The Okinawans have earned the title "the longest-lived people on earth." Fraudulent claims have been made about the Okinawa diet, attributing their extra longevity to the consumption of seafood, fruits, and vegetables. The fact that the Okinawans consumed a lot of pork, lard, and saturated fat from coconut oil has been hidden and distorted. The entire argument regarding the Okinawans is rather mute because their average life span is greater by only 16 months.

The Okinawa Centenarian Study
http://www.okicent.org/

Rank	Country	Life Expectancy	CHD
1	Okinawa	81.2	13
2	Japan	79.9	22
3	Hong Kong	79.1	40
4	Sweden	79.0	102
8	Italy	78.3	55
10	Greece	78.1	55
18	USA	76.8	100

"**Makoto Suzuki MD PhD** is a cardiologist and geriatrician. He is Professor Emeritus and former Director of the Department of Community Medicine at the University of the Ryukyus in Okinawa, Japan. Currently, he is Director, Okinawa Research Center for Longevity Science, in Urasoe, Okinawa. He recently retired from his position as professor in the Department of Human Welfare at Okinawa International University. He is Principal Investigator of the Okinawa Centenarian Study, a Japan

Chapter 3

Ministry of Health-funded study of the world's healthiest and longest-lived people. The study is entering its 31st year and is the longest continuously running centenarian study in the world."

Okinawa is a small mountainous island only 60 miles (96.5 km) long and 9 miles (14.5 km) wide in the western Pacific Ocean south of Japan. It has not been heavily industrialized, and the economy and people are rather poor compared to the large industrialized nations. These features have aided in the longevity of the people for the following reasons:

- Their weather comes from the vast Pacific Ocean and provides clean, fresh air, and the country has relatively low industrial pollution.
- The ocean provides easy access to seafood.
- The mountain people raised pigs as the primary meat supply because pigs are the most productive farm animals, and their religion did not forbid the consumption of pork and saturated fats.
- The mountains prohibited the growing of vast areas of grains, such as wheat, corn, and rice.
- The economy did not provide sufficient income for the people to import foreign foods, such as sugar, grains, vegetable oils, fruits, fruit juices, and potatoes.

The Okinawians ate a diet of meat from pigs, chickens, and fresh seafood while avoiding high-carbohydrate, high-fiber foods, such as whole grains, legumes, starchy vegetables, and sugars. They also avoided bad omega-6 vegetable oils and consumed large quantities of fresh animal fats and coconut oil. Okinawans include pork and lard in almost every dish and are proud of the fact that they "use every part of the pig except the squeal."

Top Ten Nutritional Myths

The Swedish people have a high level of heart disease which is erroneously blamed on dairy products, particularly butter. The real cause of their heart disease rate is obviously sugar. They consume large amounts of sugar in the form of sugar-laden desserts. Even so, they ranked fourth in longevity in the Okinawa Centenarian Study but had the highest coronary heart disease rate (CHD) of the seven countries listed.

The United States ranks 18th in the Okinawa Centenarian Study with a life expectancy of 76.8 years and the baseline coronary heart disease rating of 100. This is no surprise considering all the handicaps that Okinawa does not have. The United States has the following disadvantages:

- Burdened with the *USDA Food Guide Pyramid.*
- Highly industrialized with people living in large cities.
- Expansive plains ideally suited for growing grains.
- Expansive farmland ideally suited for growing potatoes.
- Expansive farmland for growing sugar beets.
- Mega-food manufacturing plants to turn grains and potatoes into hundreds of different packaged foods.
- Expensive seafood.
- Inexpensive pork but the people are told to not eat it.
- Medium-priced beef but the people are told to not eat it.

I must pause to praise the medical industry for keeping us alive as long as they do. Those involved in treating heart disease deserve a special word of appreciation. The combination of skilled doctors, excellent pharmaceutical drugs, and high-tech medical equipment has achieved wonders in treating heart disease. Heart arteries restricted with plaque can be opened in a matter of minutes with a *catheter angiography* and the placement of an artery stent. Highly-effective drugs can keep the

Chapter 3

arteries clear and can reverse heart disease when the correct treatment is followed.

Don't believe the lies propagated about saturated fat. Saturated fats actually heal the body and are highly resistant to oxidation and attack from free radicals. Saturated fats help prevent diabetes, heart disease, cancer, and bowel diseases.

People buy lean cuts of meat and avoid the fatty cuts like rib-eye steaks because they are brainwashed into thinking fat is unhealthy. It is intuitively obvious and has been proven from the study of primitive tribes that humans and carnivorous animals prefer and eat the fattiest meat. Rib-eye steaks are delicious, but lean round steak requires a high-fat dressing to make it palatable. As described in Myth No. 10 below, most dressings contain unhealthy omega-6 fatty acids rather than healthy animal fats.

Ketogenic diet with 80% fat shrinks cancer tumors.

High-Fat Ketogenic Diet Shrinks Cancer Tumors.

The following study shows that a ketogenic diet containing 80% fat shrinks cancer tumors, but a reintroduction of carbohydrates causes the tumors to grow again. Cancer research has been fully funded for 50 years without discovering the simple facts revealed in this study. Professional nutritionists say that red meat and saturated animal fats cause cancer but just the opposite is true. We have been brainwashed again.

"Metabolic management of glioblastoma multiforme using standard therapy together with a restricted ketogenic diet." Case Report - 22 April 2010 Nutrition & Metabolism 2010, 7:33doi:10.1186/1743-7075-7-33

Top Ten Nutritional Myths

Background

"Management of glioblastoma multiforme (GBM) has been difficult using standard therapy (radiation with temozolomide chemotherapy). The ketogenic diet is used commonly to treat refractory epilepsy in children and, when administered in restricted amounts, can also target energy metabolism in brain tumors. We report the case of a 65-year-old woman who presented with progressive memory loss, chronic headaches, nausea, and a right hemisphere multi-centric tumor seen with magnetic resonance imaging (MRI). Following incomplete surgical resection, the patient was diagnosed with glioblastoma multiforme expressing hypermethylation of the MGMT gene promoter."

Methods

"Prior to initiation of the standard therapy, the patient conducted water-only therapeutic fasting and a restricted 4:1 (fat: carbohydrate + protein) ketogenic diet that delivered about 600 kcal/day. The patient also received the restricted ketogenic diet concomitantly during the standard treatment period. The diet was supplemented with vitamins and minerals. Steroid medication (dexamethasone) was removed during the course of the treatment. The patient was followed using MRI and positron emission tomography with fluoro-deoxy-glucose (FDG-PET)."

Results

"After two months treatment, the patient's body weight was reduced by about 20% and no discernable brain tumor tissue was detected using either FDG-PET or MRI

Chapter 3

imaging. Biomarker changes showed reduced levels of blood glucose and elevated levels of urinary ketones. MRI evidence of tumor recurrence was found 10 weeks after suspension of strict diet therapy."

Conclusion

"This is the first report of confirmed GBM treated with standard therapy together with a restricted ketogenic diet. As rapid regression of GBM is rare in older patients following incomplete surgical resection and standard therapy alone, the response observed in this case could result in part from the action of the calorie restricted ketogenic diet. Further studies are needed to evaluate the efficacy of restricted ketogenic diets, administered alone or together with standard treatment, as a therapy for GBM and possibly other malignant brain tumors."

Gas Grilled Lamb Riblets with Tomato and Pepper Carlson's® Lemon Flavored Cod Liver Oil. Don't worry about the taste of this cod liver oil. My four-year-old grandchild said, "More, Papa."

Top Ten Nutritional Myths

Myth No. 2

Carbohydrates Are Healthy and Needed for Energy

This myth that carbohydrates are essential in the diet is stated repeatedly in health and diet articles, books, and on websites. High profile nutritionists and doctors claim that carbohydrates are necessary to provide the body with needed energy. However, the scientific fact that carbohydrates are not essential in the human diet has been known for decades. People who go on the low-carbohydrate diet quickly prove the fallacy of this myth by experiencing a big surge in energy because they are not eating carbohydrates. The lower the amount of carbohydrates in the diet, the healthier we become. Carbohydrates cause many diseases which take years to develop and have come to be called "age related"—a convenient excuse.

The anti-meat vegetarians are trying to justify eating unhealthy carbohydrates by saying, "There are bad carbohydrates and good carbohydrates." This is untrue. All carbohydrates are bad, but simple carbohydrates such as sugar and white flour are killers. Complex carbohydrates also produce debilitating age-related diseases. It just takes longer. There are no healthy carbohydrates.

> **To say there are bad carbs and good carbs is a lie. Some carbs are bad and the rest are dreadful.**

The standard recommendation states that 60% of calories in the diet should come from carbohydrates. Human anatomy and physiology textbooks may accurately state that carbohydrates are not scientifically essential to human life, but they erroneously stress that they are highly recommended for good health. The *USDA Food Guide Pyramid* recommends a high level of

Chapter 3

carbohydrates in the diet. The sweetener and grain lobbies are powerful organizations whose members infiltrate universities, government, and nutritional organizations to ensure that their high-carbohydrate products are strongly recommended. They twist and distort scientific studies to justify their products.

All of the claims that carbohydrates are scientifically required by humans—or that they are necessary for energy or good health—are absolutely false. The body can easily create the required amount of glucose from protein and fat without creating the insulin rush that occurs when carbohydrates are consumed. For example, after I had been on a highly-restrictive low-carbohydrate diet for several months, I scheduled a blood test that required a preparatory 12-hour fast. The test showed a perfect blood glucose level of 92 MG/DL, midrange in the desirable range of 65 to 109 MG/DL. The common claims that carbohydrates are required in the diet to maintain a proper blood glucose level is a myth. Eating carbohydrates places the body in a continual blood sugar control crisis that is very unhealthy.

It is distressing that highly-educated doctors and nutritionists still insist that a healthy diet requires a high level of carbohydrates. In *Life Without Bread,* Christian B. Allan, Ph.D., and Wolfgang Lutz, M.D., label carbohydrates as pathogenic. Carbohydrates cause flatulence when the intestinal bacteria feed on the undigested carbohydrates, and some of the byproducts are toxic. The pathogenic bacteria flourish in the colon because their favorite foods are plentiful—carbohydrates and fiber. The colon resists the presence of undigested carbohydrates and triggers a constipation reaction in an attempt to keep the carbohydrates from entering. Over time, the colon reverses this defense measure and triggers diarrhea in an attempt to flush the undigested carbohydrates from the body. The excessive

Top Ten Nutritional Myths

consumption of carbohydrates leads to a whole host of intestinal and autoimmune diseases, such as ulcerative colitis, asthma, and rheumatoid arthritis.

College textbooks are ideal sources for the false claims made about carbohydrates in the diet. For example, *Lippincott's Illustrated Reviews: Biochemistry* by Pamela C. Champe and Richard A. Harvey, Ph.D., claims that dietary carbohydrates are "protein sparing," which means the body will cannibalize muscle protein when carbohydrates are not consumed. This concept is absolutely ridiculous since many body builders use a low-carbohydrate diet—called the *Ketogenic Diet*—to build highly-muscular bodies. The high-protein, high-fat, low-carbohydrate, ketogenic diet is also called the *Anabolic Diet* because it puts the body into a state of building up muscles and other body structures. All body-building diets are very high-protein diets with various levels of carbohydrate restrictions.

The vegetarians of Southern India eat a low-calorie diet very high in carbohydrates and low in protein and fat. They have the shortest life span of any society on Earth, and their bodies have an extremely low muscle mass. They are weak and frail, and the children clearly exhibit a failure to thrive. Their heart disease rate discussed in the article, "Vegetarianism: An anthropological/nutritional evaluation," is double that of the meat-eaters in Northern India. In his Journal of Applied Nutrition, 1980, 32:2:53-87, H. Leon Abrams, Jr., MA, EDS, claimed the optimal diet for humans can be determined by anthropological research studies and showed that humans have primarily been meat-eaters.

Chapter 3

Carbohydrates Cause Obesity

A diet high in refined carbohydrates has been directly linked to obesity and disease. Obesity has been shown to greatly increase the risk of diabetes, cancer, heart disease, and intestinal diseases. The high-carbohydrate diet is widely accepted by the public because they have been brainwashed with the false claim that dietary fats cause obesity. The low-fat, low-cholesterol mania propagated for the last 40 years has resulted in the current obesity epidemic. The public are blamed for the explosion in obesity and are accused of not getting enough exercise. One hundred years ago, people were thin and never ran unless it was an emergency. Jogging would have been looked upon as ridiculous, unnecessary work. Their high-fat diet contained generous portions of pork lard, beef suet, chicken fat, and butter but was relatively low in carbohydrates since sugar was scarce and fruit was seasonal. Grain products in boxes (cereals) were nonexistent. Breakfast consisted of fried eggs and meat that would register on a food analysis at 70% fat on a calorie basis. This is the same diet promoted in this book.

Carbohydrates Cause Diabetes

Carbohydrates have been solidly linked to diabetes because the body reduces all carbohydrates to simple blood glucose. The excess glucose produces high levels of insulin which lead to insulin resistance in the cells and failure of the pancreas. Insulin resistance was once blamed on dietary saturated fat. This is absolutely false. Primitive people who ate an all-meat diet with the saturated animal fats never developed diabetes. Diabetics suffer a much higher incidence of heart disease, cancer, inflammatory bowel disease, and osteoporosis. A death certificate rarely lists the cause of death as diabetes. Statistics simply do not record the impact of diabetes on society. Eighty-five percent of diabetics die from heart disease and about 10% from cancer.

The constant onslaught of propaganda and lies in support of complex carbohydrates and against saturated animal fats has brainwashed society into eating more and more complex carbohydrates. The current obesity and diabetes epidemics continue to increase as more people comply with the propaganda. The meat departments in supermarkets continue to shrink as the obese, crippled shoppers motor by in battery-powered shopping carts loaded with whole grain cereal boxes, whole grain bagels, fresh fruits, 100% fruit juices, and milk.

Carbohydrates Cause Intestinal Diseases

Complex carbohydrates are not healthy as is often claimed. They appear to be healthier for the hypoglycemic because they do not cause the sharp glucose and insulin spikes, but they lead to such bowel diseases as Crohn's and ulcerative colitis. Complex carbohydrates only delay the onslaught of full-blown diabetes for a few years. Government health departments have been calling for people to eat less dietary fats and more complex carbohydrates, but this is not the answer for our current surging diabetes and bowel disease epidemics. These diseases have tripled or quadrupled since the start of the low-fat diet craze 40 years ago.

Carbohydrates Cause Osteoporosis, Bone Loss, Hip Fractures, and Degenerative Disc Disease

Carbohydrates raise the insulin hormone level to an extreme as people become insulin resistant and hypoglycemic from eating too many carbohydrates. The body reacts by increasing the levels of cortisol and adrenalin hormones, both of which demineralize

Chapter 3

the skeleton. This fact is hard science—not emotional brainwashing.

Carbohydrates knock out the joints of the skeleton. The tissues in the joints are made from amino acids and fatty acids, and both are displaced in a diet containing whole grains, fruits, and vegetables. As a result, physicians now specialize in hip and knee joint replacements.

Medical reports seldom explain the importance of amino acids in bone health. The strength of bone is achieved by the collagen matrix which is composed of protein molecules derived from amino acids. Meat is the best food for a complete assortment of essential amino acids necessary for building thousands of different protein molecules in the body. Food and vitamin salesmen are telling women to consume more calcium and yogurt for strong bones instead of meat. They direct this propaganda toward women because women have a higher incidence of hip fractures. Women also shun red met more than men do. The connection is perfectly clear without a double-blind study. The collagen matrix in bone functions in the same manner as steel reinforcement bar in concrete. The mineral matrix in bone is not simply calcium. Magnesium and other minerals are vitally important. The mineral complex functions in the same manner as cement, sand, and gravel in concrete. Together they provide a strong structure. The collagen matrix resists the tension stress, and the mineral matrix resists the compressive stress. Hip fractures almost always occur because the collagen matrix snaps. Broken bones are rarely compression fractures of the mineral matrix. The absolute scientific fact is that hip fractures occur because of a protein deficiency—not a calcium deficiency.

Top Ten Nutritional Myths

Stress fractures occur in vertebra because the collagen matrix has been cannibalized by cortisol and adrenaline. Twenty years on a high-carbohydrate, low-protein diet will devastate your skeleton.

> **Protein Deficiency Causes Hip Fractures, Not a Calcium Deficiency.**

Carbohydrates Cause Cancer

Carbohydrates increase the risk of cancer by producing intestinal diseases, reducing the effectiveness of the immune system by the stimulation of cortisol hormone, and generally creating degenerative diseases. Carbohydrates in the diet displace protein and fat. The immune cells in the body needed to fight cancer are all made from polypeptides—complex molecules of amino acids. Meat is the best food for the entire complex of amino acids needed to build immune system cells. The major types of immune cells are:

- B cells and T cells of the lymphocytes type that mark a target with receptor molecules.
- Killer T cells are a sub-group of the T cells that attack infected cells that have been marked.
- Helper T cells regulate the immune system's response to recognized pathogens.

Meat in the diet provides all of the amino acids needed to build strong immune system cells that the body uses to destroy cancer cells. During their two explorations between the years 1906 and 1912, Arctic explorers Vilhjalmur Stefansson and Karsten Anderson never observed a single case of cancer among the Eskimos who ate a 100% meat diet.

Chapter 3

Carbohydrates Cause Heart Disease

Carbohydrates cause heart disease by raising the blood level of the bad triglycerides and lowering the level of the good HDL (high density lipoproteins). Diabetics are good examples. Heart disease kills diabetics at a rate of 85%, which is much greater than those without diabetes.

> "High triglyceride levels, in conjunction with high LDL and low HDL, significantly increase the risk for developing cardiovascular disease. Controlled carbohydrate nutrition effectively lowers triglycerides while raising HDL cholesterol. Lowering the LDL to HDL ratio in conjunction with lowering triglycerides will significantly reduce the risk of coronary heart disease." Dr. Robert C. Atkins.

High levels of blood glucose in the presence hyperinsulinemia cause polysaccharide chains to attach to serum proteins, artery proteins, and LDL molecules in a process called *glycation*. LDL and glycated protein form a fatty deposit called *atherotic plaque*. As a result, the serum molecules and the fatty deposits become sticky. White blood cells called *macrophages* ingest the glycated proteins and LDL. This causes the macrophages to swell to form foam cells which attach to the sticky artery walls to narrow the artery and restrict blood flow. These conditions are produced by a diet high in carbohydrates. The process occurs rather suddenly in people who are hyperglycemic, hypoglycemic, or diabetic.

In a relatively short period of time, clear arteries can become plugged and lead to a heart attack. Arteries have become plugged within a year in those patients who have had bypass surgery.

Top Ten Nutritional Myths

Carbohydrates Cause Dental and Bone Diseases

Tooth decay (caries) occurs when plaque bacteria turn carbohydrates into acids that attack tooth enamel. Carbohydrates feed bacteria in the mouth where the weakened enamel allows entrance. This combination eventually results in cavities. Carbohydrates cause the dental arch to be narrow which crowds teeth and makes them come in crooked. Carbohydrates cause the pelvic arch to be narrow and make childbirth more difficult. All of these conditions were solidly proved nearly 80 years ago by dentist Weston A. Price, DDS, who traveled the world with his wife in the 1920s and 1930s to study the health of isolated people and tribes. He found that all those who lived on a low-carbohydrate diet had excellent dental health and bone formation. His studies are well documented in *Nutrition and Physical Degeneration*.

The damage caused to bones and joints by a high-carbohydrate diet was discussed previously but is worth repeating. Repetition may help to remove some of the brainwashing that has swamped the nutrition/health debate for the last 60 years.

Vegetarians have falsely claimed that a high-protein diet causes calcium to be leached from the bones and excreted in the urine. This is absolutely false. Calcium in the body can be harmful. Calcium deposits in muscle tissue cause a painful, irreversible condition. Calcium causes hardening of the arteries, a deadly condition. The high-carbohydrate diet leads to high insulin, which in turn leads to high cortisol. The cortisol causes osteoporosis when the body extracts the high tensile strength collagen fibers from bones, removes the mineral matrix by demineralization, and weakens connective tissues at the joints. Cortisol literally causes people to pee away their bones. The result is a rapid onset of osteoporosis, degenerative disc disease,

Chapter 3

and the high risk of a hip fracture. I have described this sudden onset of osteoporosis as the *Rieske Instant Osteoporosis Cycle (IOC)*.

Carbohydrates Cause Eye Diseases

High insulin levels cause eye diseases. Cataracts may be the result of high insulin and the high-carbohydrate diet as outlined on the *USDA Food Guide Pyramid*. Cataracts may not be genetic as claimed. Eating pasta, cereal, bread, and cakes is genetic because people cook and eat the same foods as their parents. Carbohydrates cause people to develop diabetes, and the diabetic has a high risk of developing retinitis pigmentosa, macular degeneration, and optic nerve damage, all of which may lead to blindness.

Carbohydrates in the Form of Starches Damage the Liver

> "A diet rich in potatoes, white bread and white rice may be contributing to a "silent epidemic" of a dangerous liver condition. "High-glycaemic" foods - rapidly digested by the body - could be causing "fatty liver", increasing the risk of serious illness. Boston-based researchers, writing in the journal Obesity, found mice fed starchy foods developed the disease. Those fed a similar quantity of other foods did not. One obesity expert said fatty liver in today's children was "a tragedy of the future". BBC News, UK, September 21, 2007.

The universally-accepted concept of a "balanced diet" is without any basis in fact or science. We are told to eat high quantities of fruit and whole grains and avoid red meat and saturated fats. A high percentage of the populations in English-speaking countries follow this dogma without the slightest question. Mankind has lived thousands of years with very little or no fruit or grain in the diet. Primitive people never had fruit available year round, and many lived in a climate where fruit was never available. Many people have a condition known as *fructose malabsorption* (previously called *fructose intolerance*) in which the body's ability to metabolize fructose is compromised. The fructose causes insulin resistance which contributes to diabetes and heart disease. Seventy-five percent of the space in the fresh fruit and vegetable sections of the typical supermarket in the USA is loaded with high-carbohydrate fruits and starchy vegetables such as potatoes and yams. Only 25% of the space is allotted to the healthy low-carbohydrate vegetables. Shopping carts can be observed loaded with fruit and products made from grains as the obese customers waddle to the checkout area with their balanced diet.

Carbohydrates Have Caused the Epidemic of Metabolic Syndrome

It is a scientific fact that fructose sugar in fruit produces insulin resistance which increases the risk of Alzheimer's disease, heart disease, high blood pressure, cancer, and diabetes. See the following truthful study which proves that fructose (found in abundance in fruit and fruit juice) is one of the root causes for metabolic syndrome, diabetes, heart disease, cancer, inflammatory bowel disease, and Alzheimer's disease.

"Hypothesis: fructose-induced hyperuricemia as a causal mechanism for the epidemic of the metabolic syndrome."
http://www.ncbi.nlm.nih.gov/pubmed/16932373?dopt=Abstract Plus

> "The increasing incidence of obesity and the metabolic syndrome over the past two decades has coincided with a marked increase in total fructose intake. Fructose--unlike

Chapter 3

other sugars--causes serum uric acid levels to rise rapidly. We recently reported that uric acid reduces levels of endothelial nitric oxide (NO), a key mediator of insulin action. NO increases blood flow to skeletal muscle and enhances glucose uptake. Animals deficient in endothelial NO develop insulin resistance and other features of the metabolic syndrome. As such, we propose that the epidemic of the metabolic syndrome is due in part to fructose-induced hyperuricemia that reduces endothelial NO levels and induces insulin resistance. Consistent with this hypothesis is the observation that changes in mean uric acid levels correlate with the increasing prevalence of metabolic syndrome in the US and developing countries. In addition, we observed that a serum uric acid level above 5.5 mg/dl independently predicted the development of hyperinsulinemia at both 6 and 12 months in nondiabetic patients with first-time myocardial infarction. Fructose-induced hyperuricemia results in endothelial dysfunction and insulin resistance, and might be a novel causal mechanism of the metabolic syndrome. Studies in humans should be performed to address whether lowering uric acid levels will help to prevent this condition."

Are fruits and fruit juices wholesome and healthy foods as proclaimed my most professional nutritionists? Absolutely not!

The "Balanced Diet" Is Very High in Carbohydrates

The "balanced" *Mediterranean Diet* is another big scam. Obesity is as rampant among the older women of Greece and Italy as it is in all English-speaking countries. The diet does not stand up to close investigation. The people of Greece and Italy are suffering an epidemic of obesity, diabetes, heart disease, and cancer. The

Top Ten Nutritional Myths

health claims for the diet are all scams supported by book sellers, olive oil salesmen, and animal rights advocates—people who are against eating red meat. The real test is to feed the Mediterranean Diet to elderly people who already have diabetes, heart disease, Crohn's disease, and ulcerative colitis. It will not help them. In fact, it will make them worse.

> **Obesity and disease are increasing as more diet and health information is made available because people believe the "balanced-diet" lies.**

Grains are a new human food. Many people are gluten intolerant or have other grain allergies that lead to bowel diseases. The balanced diet concept is a myth that will harm your health rather than make it better. The current epidemic of cancer, heart disease, diabetes, and bowel diseases has proved the balanced diet of the *USDA Food Guide Pyramid* will make you sick and shorten your life. The balanced diet concept is scientifically false.

**High-Fat 73% / 27% Beef Burger with
Swiss Cheese, Mushrooms, and Asparagus**

Chapter 3

Myth No. 3

Fiber Is Healthy and a Requirement in the Diet

The myth that fiber is required in the diet is so widespread and universally accepted that even many low-carbohydrate proponents believe the lie without hesitation. The fiber theory was generated by the low-fat, high-carbohydrate diet supporters to combat the symptoms of constipation and diarrhea caused by eating carbohydrates. They recommend a high-fiber diet as the solution for both constipation and diarrhea, but these symptoms are actually the result of an excessive consumption of carbohydrates. The fact that fiber increases mobility of the material in the intestines is seen as positive; however, much of the bulk in the stool is live and dead bacteria which are produced in mass by the carbohydrate and fiber. The high-carbohydrate diet promotes the overgrowth of bacteria and yeasts which becomes a serious problem as mobility slows. Many of these are unhealthy pathogenic bacteria that cause irritation of the intestines and eventually lead to "leaky gut syndrome" and bowel diseases as well as a host of other autoimmune diseases. Fiber is cellulose that cannot be digested as a human food because we lack the cellulace enzyme; however, bacteria feast on fiber, especially in the colon where they wreck havoc.

Fiber is not required in a low-carbohydrate diet because intestinal bacteria are restricted by the lack of a food source—carbohydrates. High-fiber foods, such as grains, seeds, and fruit are also very high in carbohydrates. Grains make up the highest percentage of raw materials used for the manufacture of high-carbohydrate foods. The grain lobby is very powerful in promoting their products, and their resources to do so are almost limitless. They claim fiber reduces diabetes, cancer, heart

Top Ten Nutritional Myths

disease, and intestinal diseases, but it is a well-proven fact that high-carbohydrate diets cause these diseases.

A study of ancient societies who lived on a high-fiber, high-carbohydrate diet easily proves the unhealthy effects. Ancient Egyptians are perfect examples. Their diet was based on a high percentage of whole grains, fruits, and vegetables in which the fiber content was very high. The diet was low in fat, and they did not eat refined sugars. These Egyptians of the Pharaohs' times ate a highly vegetarian diet. The results were disastrous. Their writings and the study of mummies reveal they had a high incidence of diabetes, heart disease, intestinal diseases, arthritis, osteoporosis, and poor dental health. Their high-fiber diet—which had no refined carbohydrates—did not produce the good health as promised by the majority of our modern dietary references and professional medical and nutritional associations. The tens of thousands of well-preserved Egyptian mummies give us the absolutely solid scientific proof that the high-fiber, high-carbohydrate diet is very unhealthy.

The common thought that it is necessary to have several bowel movements each day is a myth. The low-fat, low-protein, high-carbohydrate diet usually produces several bowel movements per day. Chronic constipation and diarrhea are common body reactions of the digestive system to the constant bombardment by fiber and carbohydrates. Sufferers attempt to correct the condition by taking fiber supplements as recommended by their doctor only to discover they are not effective. The low-carbohydrate, low-fiber diet will produce normal stools with bowel movements that sometimes skip two days and frequently skip one day. The intestinal tract becomes calm and bowel movements are natural. One should not expect several bowel movements per day which typically occur on the disease-causing high-fiber, high-carbohydrate diet.

Chapter 3

Fiber absorbs vitamins and minerals and discharges them from the body. Calcium and magnesium are desperately needed to prevent osteoporosis, bone loss, hip fractures, and degenerative disc disease, but fiber leaches these minerals from the digestive tract and discharges them in the stool.

Do not take fiber supplements. Do not take psyllium seed husks which are very abrasive to the digestive system. Do not eat wheat bran or rice bran. Fiber is a bad dude. While dietary fiber may not be digestible by the healthy individual, it certainly is digestible by pathogenic intestinal bacteria and yeasts. Fiber is the perfect time-release food for bad colon bacteria and yeasts. It is one of the worst possible foods. These bacteria and yeasts ferment the fiber to produce alcohol, acetaldehyde, lactic acid, acetic acid, and a host of other toxic chemicals as they break down the fiber.

> **Fiber is the perfect time-release food for bad colon bacteria and yeasts. It is one of the worst foods you could eat.**

Intestinal gas is a sure sign that fiber and/or sugars are being fermented. The vegetarian concept of turning the intestines into a fermentation tube is ridiculous. Avoid all whole grains, brown rice, fruit, and dried beans as they are high in both fiber and complex carbohydrates, a double blow to the digestive system. Many gastroenterologists recommend a high-fiber diet based on the faulty logic of Dr. Dennis Burkitt, a British surgeon who worked in Africa more than half a century ago. Dr. Burkitt's theory that barley bread prevented irritable bowel disorders was seriously flawed. The Africans were simply showing the benefits of not eating fruit and refined carbohydrates including sugar and flour. Since their barley grain was probably not ground very well,

Top Ten Nutritional Myths

the fiber was difficult for the intestinal bacteria to attack; this saved them from the health hazards of eating fiber. Our finely ground grains of today do not produce the same result. Fiber not only does not prevent or cure irritable bowel diseases, it actually makes them worse. In *Nutrition and Physical Degeneration,* Weston A. Price, DDS, described primitive people living in excellent health on a low-fiber diet. During their many years of living with the Eskimos, Arctic explorers Vilhjalmur Stefansson and Karsten Anderson proved that a very low-fiber diet prevents intestinal diseases and cancer. Most current doctors and nutritionists simply ignore these findings that a zero-fiber diet produces excellent health and prevents colon cancer. New scientific studies that indicate fiber may raise the risk of colon cancer are also ignored. Fiber does not prevent cancer as erroneously claimed by most modern-day professionals.

Dietary Fiber and the Risk of Colorectal Cancer and Adenoma in Women

> "Background: A high intake of dietary fiber has been thought to reduce the risk of colorectal cancer and adenoma."

> "Conclusions: Our data do not support the existence of an important protective effect of dietary fiber against colorectal cancer or adenoma." (New England Journal of Medicine, January 21, 1999; 340:169-76.)

Did this study say fiber does not protect against colorectal cancer or adenoma? Yes, it does.

Do not fall for the "colon cleanse and detoxification" scam. The colon does not contain a lining of putrid material that looks like "chunks of debris that resemble cooked liver, long black twisted rope-like pieces." A colon cleanse or a detox program using harsh

Chapter 3

herbs and fiber products only serves to create more problems that may lead to leaky gut syndrome.

The US Food and Drug Administration (FDA) has established a Recommended Daily Allowance (RDA) of 25 grams of fiber per day; yet there is no scientific basis for this value, and research studies on fiber are scant. No studies support the requirement for fiber, but many expose health hazards associated with fiber. The low-carbohydrate diet recommended in this book contains three grams of fiber or less from non-starchy vegetables.

The Eskimos observed by Arctic explorer Vilhjalmur Stefansson clearly proved that eating a diet totally devoid of fiber is perfectly healthy. In fact, it is healthier than eating a diet with fiber. The evidence is presented in Stefansson's book, *The Fat of the Land.*

Vegetarians strongly advocate fiber because of its high content in grains, legumes, and fruit, and its low content in meat. *The Bran Wagon* by Barry Groves, PhD, exposes many of the common myths about high-fiber foods. Dietary fiber actually causes or increases the severity of many diseases.

Low-Fiber Fish Dinner

Top Ten Nutritional Myths

Myth No. 4

Red Meat, Such as Beef, Lamb, and Pork Is Unhealthy

The claims that red meat, such as beef, lamb, and pork, is unhealthy are **myths** with absolutely no basis in scientific studies or logical nutritional data. The reverse is actually true. Red meat with its natural fat is a very healthy food and heals the body of many diseases. Red meats do not contain unnatural levels of hormones as stated by the numerous sources. Any hormone given to steers is eliminated naturally by their bodies well before slaughter. These meats do not contain unhealthy levels of pesticides. Any pesticide that may have been on the animals' food or sprayed on their bodies to control pests is also eliminated naturally before slaughter. Laws and meat inspections prevent the sale of red meat with any unhealthy levels of hormones or pesticides. You have been brainwashed if you believe otherwise.

Our digestive organs and digestive enzymes are designed to show that mankind is basically a carnivorous (meat eating) species with the ability to digest carbohydrates from fruits and vegetables. Health is damaged when carbohydrates make up a great percentage of our diet. Meat does not putrefy in the intestines as falsely claimed by vegetarians. The stomach secretes a large amount of acid and enzymes to break down the protein into individual amino acids for absorption in the small intestines. Your body was designed to digest meat and does so with the aid of "good" bacteria in the intestines—often referred to as *probiotics* (for life).

The North American Plains Indian is a prime example of a society that lived for centuries on the red meat of the buffalo. The Indians made pemmican, which was a combination of dried,

Chapter 3

crushed, and shredded buffalo meat mixed with heated buffalo fat at a 50% rate. The fat content was about 70% to 80% on a calorie basis. Protein was 20% to 30% with no carbohydrates or fiber. There is no documented data that this primarily red-meat diet had any adverse health effects on the Indians. The opposite is true. They were extremely healthy until the white man gave them high-carbohydrate foods. Now the Pima Indians of Arizona have the highest diabetes rate in the world and a very high level of obesity. Many Indians are returning to the red meat pemmican of their ancestors in order to regain their health.

Awesome health and a robust immune system are the keys to cancer prevention.

Highly-processed meats have been linked to colon cancer. Early results of a major new study suggest that eating lots of preserved meats, including salami, bacon, cured ham, and hot dogs, could increase the risk of bowel cancer by 50%. When it came to fresh red meat—beef, lamb, pork, and veal—there was no link. Previous studies have linked a high meat intake to colorectal cancer; however, these studies grouped fresh and processed meats together, which produced a distorted result.

Cancer is not caused by eating animal products and natural saturated animal fats.

Nutritional books and reference resources consistently understate or neglect to state the high vitamin and mineral content of meat. Fruits and whole grains are generally listed as the first sources in order to give the impression that they are preferred. Primitive Indian and Eskimo tribes have proved that a meat-only diet is healthy and will also cure many diseases.

Top Ten Nutritional Myths

Vitamin B-12 is only available from animal sources, so vegetarians are at a constant risk of deficiency.

Red meat does not cause or contribute to diabetes, heart disease, and cancer as claimed or advocated by its opponents.

Cancer is caused by the combination of free radical damage to normal cells and the lack of body-building amino acids, fatty acids, vitamins, and minerals. A growing number of professionals believe that another major cause of cancer is the excessive consumption of carbohydrates in the form of sugars, grains, fruits, and starchy vegetables. The third major cause for cancer is believed to be the consumption of hydrogenated vegetable oils (trans fats) in the diet.

Meat and natural fats are just the foods which promote health and prevent cancer. The cancer rate has risen in all populations with an increased consumption of carbohydrates. Our modern diet—as outlined on the *USDA Food Guide Pyramid* and other health organizations—has replaced the healthy, natural animal fats with carbohydrates and unhealthy omega-6 vegetable oils. The result has been a very high level of cancer even though many have stopped smoking. The harsh criticisms of red meat and saturated fat are not supported by scientific research or anthropological surveys.

Eating meat has been associated with lower blood pressure in a study of Chinese men and women.

> Int J Epidemiol 2002 Feb;31(1):227-33
>
> "Inverse relationship between urinary markers of animal protein intake and blood pressure in Chinese: results from the WHO Cardiovascular Diseases and Alimentary Comparison (CARDIAC) Study."

Chapter 3

"CONCLUSION: The results provide strong evidence that animal protein intake is associated inversely with BP in Chinese populations."

The low intake of saturated fats and meat has been linked to an increase of stroke (intraparenchymal hemorrhage) in women according to the following study:

Circulation 2001 Feb 13;103(6):856-63

"Prospective study of fat and protein intake and risk of intraparenchymal hemorrhage in women."

"CONCLUSIONS: Low intake of saturated fat and animal protein was associated with an increased risk of intraparenchymal hemorrhage, which may help to explain the high rate of this stroke subtype in Asian countries. The increased risk with low intake of saturated fat and trans unsaturated fat is compatible with the reported association between low serum total cholesterol and risk."

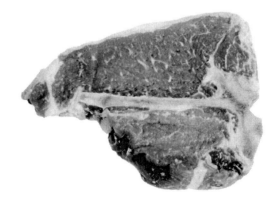

**Porterhouse or T-Bone Steak
One of the Best Foods You Could Eat**

Top Ten Nutritional Myths

Myth No. 5

Organic Fruits, Vegetables, Eggs, and Meat Are Healthier than Nonorganic Food

Organic fruits and vegetables are more likely to contain bacterial contamination than conventionally-grown foods and are therefore less healthy. The sale of organic fruits and vegetables is a scam because many of these products can only be grown using pesticides. Those big, beautiful, perfectly-formed "organic" apples in the store are a fraud because worm-free apples are absolutely impossible to grow without pesticides. Organic producers are driven by outrageous profit margins to support their propaganda.

Worm-free organic apples are impossible to grow.

> "Researchers are still looking for the first human death from pesticide residues, fifty years after DDT was introduced and thirty years after its use was banned in the United States, but manure is apparently claiming lives almost daily through bacterial contamination of organic food."
>
> "What most people don't realize - and activists try to hide - is that organic farmers are allowed to use a wide array of natural chemicals as pest killers. Moreover, these natural poisons pose the same theoretical (but remote) dangers as the synthetic pesticides so hated by organic devotees." *The Hidden Dangers In Organic Food* by Dennis T. Avery.

Imported meats and most of our winter fruits and vegetables come from foreign countries in Central America or the Southern Hemisphere. Some come from local hothouses. Illegal drug plants

Chapter 3

can be growing in one field and animals, fruits, or vegetables in another. They will quickly sell their meat and produce with the fraudulent claim that it is "organically grown" in order to get the larger premium price. Nothing could be farther from the truth. Not only are they using pesticides, antibiotics, and hormones to get the most productivity, but many products have been found to be contaminated with chemicals which are illegal in all English-speaking countries. The word *organic* is almost meaningless because of cheating in the food industry. Buying organic products is a foolish waste of money that only serves to encourage and benefit the cheaters.

No studies have proved that organic fruits and vegetables provide any health benefit. This is pure conjecture based on the assumption that non-organic foods contain unhealthy levels of pesticides or herbicides. To the contrary, studies have consistently shown no discernable health difference between organic and non-organic foods.

Organic foods do not taste better than conventionally grown foods. Taste is a matter of freshness and allowing the food to ripen before harvest.

Organic fruits and vegetables are much more likely to be contaminated with unhealthy Escherichia coli (E. coli) and other bacteria. Many deaths have been traced to fruits and vegetables from an outbreak of E. coli bacteria. Conventionally grown vegetables have been shown to be healthier because they contain less bacterial contamination.

Many people have become so paranoid about pesticides that they think they will surely develop cancer if they eat conventionally farmed produce. This is a myth because present-day pesticides are tested extensively for safety. As a teenager on a farm, I was

Top Ten Nutritional Myths

literally drenched in pesticides, including DDT®, Malathion®, and two others. My father put four different pesticides in the tank of a heavy-duty sprayer that I pulled with the tractor. The sprayer had a 1" (25 mm) diameter hose and a pump discharge pressure of 400 pounds per square inch (27.6 bars) that would literally drench a full-grown apple tree in seconds. The high-pressure, high-volume spray created a cloudy mist of overspray which dried to form a pure white coating of pesticides as the moisture evaporated. The pesticide remained on my arms, neck, and face for several hours until spraying was completed. I received more pesticide in each spraying than the average person receives in a lifetime. We sprayed several times each year for many years, but my severe overdose with four different pesticides has not affected my health in any way that I can discern—even though DDT® and Malathion® have been banned by government regulations. Maybe the pesticides have prevented cancer, diabetes, and IBS.

The myth that a "free-range" chicken egg is healthier is related to the beef scam concerning omega-3 fatty acids. The plea to eat eggs only from free-range chickens is a scam. Keeping chickens and turkeys in cages is perfectly fine. Chickens and turkeys must be the dumbest birds in the world—as those who have raised them can attest. Commercial free-range chickens may be given an access to an outside area, but it is usually nothing but barren dirt where they rarely go. The idea that they are walking around in the grass eating insects is simple ignorance. When given the chance to be free, they will return to their cage. Chickens would certainly stop laying eggs if the cage caused them to be under stress, but they lay eggs even more frequently. The omega-3 egg is also a scam. Oh, certainly they could have a tiny bit more omega-3 fatty acids if the producer is not cheating. Do not depend on eggs as your source for omega-3 fats because all eggs

Chapter 3

are deficient in omega-3 fats. It is much less expensive to buy regular eggs and eat some cold water fish or take cod liver oil as mentioned below. A reader sent the following comment about free-range chickens:

> "I agree with you the "free range" idea is bunk. I have been to a "free range" farm. It consisted of huge barns of chickens (10,000 per barn) that were meandering across concrete floors. That's what free range means here in New England."

Cows are almost as dumb as chickens. A cow kept in a beautiful grassy pasture will lie down in its own excrement. Free-range chickens in the same pasture will scratch and pick in the excrement to find lunch. Animals do not have human emotions or knowledge because they are stupid animals.

One tablespoon of cod liver oil has more omega-3 fatty acid than four dozen eggs.

Buy Carlson's® Lemon Flavored Cod Liver Oil and avoid capsules because they are often used to conceal poor grades of oil. Omega-3 fats easily become oxidized and rancid as can be identified by the smell. Food processors remove them from their products to increase shelf life and eliminate the need for refrigeration.

A shopper at the local natural foods market was observed buying half a dozen (6) brown, organic, cage free chicken eggs for $2.49. She could have purchased five dozen (60) regular eggs at the nearby warehouse store for $2.89. She paid 8.6 times, or 760% more for eggs that are no different than the regular eggs (2006 prices). The organic food craze has reached the level of mass delusion.

Top Ten Nutritional Myths

Myth No. 6

Cholesterol Causes Heart Disease

This myth may never go away. The profit motive behind specialty foods, prescription cholesterol-lowering drugs, blood testing, and doctor visits results in billions of dollars of business each year. The amount of money spent to keep this myth going is astronomical. The United States has spent hundreds of millions of dollars funding studies to support this myth and has failed to prove any connection between cholesterol and heart disease. Half of the people with heart disease have normal cholesterol levels. The only connection is among people with hypercholesterolemia, which is a rare disease of extremely high cholesterol.

There are strong indications that low HDL cholesterol is related to an increase in heart disease, and this is commonly found in people with low (not high) total cholesterol readings. Eating a low-cholesterol diet does not reduce blood cholesterol because the liver simply produces more for the transport and storage of body fat.

Low cholesterol increases the risk of cerebral hemorrhage (stroke), gallbladder disease, and many types of cancers. Two major factors that affect heart disease are lipoprotein, Lp(a) and homocysteine, neither of which is normally checked by doctors. A lipoprotein reading of 20 mg/dl is normal and above 30 mg/dl is elevated. A Homocysteine level below 13 micromoles/L is normal.

> "For almost forty years, the lipid hypothesis or diet-heart idea has dominated medical thinking about heart disease. In broad outlines, this theory proposes that when we eat foods rich in saturated fat and cholesterol, cholesterol is

Chapter 3

then deposited in our arteries in the form of plaque or atheromas that cause blockages. If the blockages become severe, or if a clot forms that cannot get past the plaque, the heart is starved of blood and a heart attack occurs.

Many distinguished scientists have pointed to serious flaws in this theory, beginning with the fact that heart disease in America has increased during the period when consumption of saturated fat has decreased. "The diet-heart idea," said the distinguished George Mann, "is the greatest scam in the history of medicine." And the chorus of dissidents continues to grow, even as this increasingly untenable theory has been applied to the whole population, starting with low-fat diets for growing children and mass medication with cholesterol-lowering drugs for adults." *What Causes Heart Disease* by Sally Fallon and Mary G. Enig, Ph.D.

"Anthropological Research Reveals Human Dietary Requirements for Optimal Health"
by H. Leon Abrams, Jr., MA, EDS

"Michael DeBakey, world renowned heart surgeon from Houston, who has devoted extensive research into the cholesterol coronary disease theory, states that out of every ten people in the United States who have atherosclerotic heart disease, only three or four of these ten have high cholesterol levels; this is approximately the identical rate of elevated cholesterol found in the general population. (10) His comment: "If you say cholesterol is the cause, how do you explain the other 60% to 70% with heart disease who don't have high cholesterol?" In 1964 DeBakey made an analysis of cholesterol levels from usual hospital laboratory testing of 1,700 patients with atherosclerotic disease and found there was no positive or definitive relationship or correlation between serum cholesterol levels and the extent or nature of atherosclerotic disease."

The recommended meal shown in this picture violates the standard low-cholesterol, low-fat diet that is making people fat.

Eggs Fried in Coconut Oil or Butter with Melted Swiss Cheese and Avocado

The scientifically-based diet that I present is intended to overwhelm the body with all of the essential nutrients and other trace elements, some of which should be classified as essential. The vitamins, minerals, and other supplements I suggest will ensure that this goal is met, especially by those suffering from digestive ailments. This book is not about a diet that simply avoids nutritional deficiencies as is the focus of many professional nutritionists.

Chapter 3

Myth No. 7

Dietary Protein Requirement Is 10% on a Calorie Basis

This statement is absolutely false even though it is the official position of the *USDA Food Guide Pyramid.* Protein—especially from red meat—is falsely accused of causing heart disease, colon cancer, and kidney disease.

Diabetes Solution - Chapter 9 by Dr. Richard K. Bernstein

> "If you are a long-standing diabetic and are frustrated with the care you've received over the years, you have probably been conditioned to think that protein is more of a poison than sugar and is the cause of kidney disease. I was conditioned the same way—many years ago, as I mentioned, I had laboratory evidence of advanced proteinuria, signifying potentially fatal kidney disease—but in this case, the conventional wisdom is just a myth."
>
> "Non-diabetics who eat a lot of protein don't get diabetic kidney disease. Diabetics with normalized blood sugars don't get diabetic kidney disease. High levels of dietary protein do not cause kidney disease in diabetics or anyone else. There is no higher incidence of kidney disease in the cattle-growing states of the United States, where many people eat steak every day, than there is in the states where beef is more expensive and consumed to a much lesser degree. Similarly, the incidence of kidney disease in vegetarians is the same as the incidence of kidney disease in non-vegetarians. It is the high blood sugar levels that are unique to diabetes, and to a much lesser degree the high levels of insulin required to cover them (causing hypertension), that cause the complications associated with diabetes."

Top Ten Nutritional Myths

The body actually uses natural unprocessed fresh beef and other natural meats to heal and prevent disease. The human immune system is composed of protein molecules made from meat, and a diet high in meat will provide a robust immune system that prevents and cures diseases. Animal protein does not cause osteoporosis, reduce bone density, or produce kidney stones as claimed by many vegetarians and others who propagate lies.

"High-Protein Diets: Separating Fact from Fiction"
Stephen Byrnes, Ph.D., RNCP

> "It is excessive carbohydrate intake, not protein or animal protein intake, that can result in heart disease and cancer (1). Readers should note that the type of diet Gilbert advocates in her article is a high-carbohydrate one because that is exactly what diets that are low in protein and fat are. Furthermore, the idea that animal products, specifically protein, cholesterol, and saturated fatty acids, somehow factor in causing atherosclerosis, stroke, and/or heart disease is a popular idea that is not supported by available data, including the field of lipid biochemistry (2)."

> "The claim that animal protein intake causes calcium loss from the bones is another popular nutritional myth that has no backing in nutritional science. The studies that supposedly showed protein to cause calcium loss in the urine were NOT done with real, whole foods, but with isolated amino acids and fractionated protein powders (3). When studies were done with people eating meat with its fat, NO calcium loss was detected in the urine, even over a long period of time (3). Other studies have confirmed that meat eating does not affect calcium balance (4) and that protein promotes stronger bones (5).

Chapter 3

Furthermore, the saturated fats that Gilbert thinks are so evil are actually required for proper calcium deposition in the bones (6)."

"The reason why the amino acids and fat-free protein powders caused calcium loss while the meat/fat did not is because protein, calcium, and minerals, require the fat-soluble vitamins A and D for their assimilation and utilization by the body. When protein is consumed without these factors, it upsets the normal biochemistry of the body and mineral loss results (7). True vitamin A and full-complex vitamin D are only found in animal fats."

"If the protein-causes-osteoporosis theory teaches us anything, it is to avoid fractionated foods (like soy protein isolate, something Gilbert would no doubt encourage readers to consume given her zeal for soy) and isolated amino acids, and to eat meat with its fat. New evidence shows that men and women who ate the most animal protein had better bone mass compared to those who avoided it (8) and that vegan diets (most likely also advocated by Gilbert) place women at a greater risk for osteoporosis (9)."

"The claim that protein intake leads to kidney stones is another popular myth that is not supported by the facts. Although protein restricted diets are helpful for people who have kidney disease, eating meat does not cause kidney problems (10). Furthermore, the fat-soluble vitamins and saturated fatty acids found in animal foods are pivotal for properly functioning kidneys (11)." Dr. Byrnes' references are not included in this book.

Top Ten Nutritional Myths

Life Without Bread
by Christian B. Allan, Ph.D., and Wolfgang Lutz, M.D.

This book illustrates how the connection between high carbohydrate consumption and cancer is becoming more evident.

> "After carbohydrates are consumed, the levels of sugar and glucose in the blood rise. The body responds by releasing insulin from the pancreas into the bloodstream. The carbohydrate theory of cancer is simple:"

> "Too much insulin and glucose in the blood can cause cells to dedifferentiate, just as they do in cell lines, and thus can be a primary cause of dietary-related cancer." Pages 169-170.

> "There have been many studies done, in animals and people, that indicate that fat content in the diet is not responsible for breast cancer or any other cancer. We know there's a tendency to blame dietary fat for just about everything that goes wrong, but that's just a lazy way out. Time after time, the studies show it just isn't true." Page 173.

> "The Eskimos who ate only fat and protein never had any cancer in their population until a Western (high-carbohydrate) diet was introduced. Why don't we ever hear of cancer of the heart? Probably because the heart uses almost all fat for energy, thus cancer does not have a chance to develop in those cells. We hope that researchers will take the next step and start looking at what has been known for a long time. Dietary related cancer is a sugar metabolism disease just like all the others." Page 177.

Chapter 3

"Compound in meat prevents diabetes, study suggests"

"West Lafayette, Ind. -- A common type of fat found in red meats and cheeses may prevent diabetes, according to a research team from Purdue University and The Pennsylvania State University."

Studies Show Eating Meat Builds Stronger Bones

A high-protein diet boosts healthy antioxidant levels—a low-protein diet induces oxidative stress. A large study published by the *Journal of Bone and Mineral Research* (1) found that elderly men and women who consumed the most animal protein had the lowest rate of bone loss while those who consumed the least amount of animal protein had the highest rate of bone loss. A study by the *American Journal of Clinical Nutrition* (2) found that postmenopausal women who consumed the highest amount of animal protein had the strongest bones and the lowest percentage of hip fractures. Diets lowest in meat resulted in a longer healing time for bone fractures and the slowest recovery time for illnesses in general. Eating meat improves healing and health.

(1) Hannan MT, Tucker KL, Dawson-Hughes B, Cupples LA, Felson DT, and Kiel DP. "Effect of Dietary Protein on Bone Loss in Elderly Men and Women: The Framingham Osteoporosis Study." *Journal of Bone and Mineral Research.* 2000;15:2504-2512.

(2) Munger RG, Cerhan JR, and Chiu BCH. "Prospective study of dietary protein intake and risk of hip fracture in postmenopausal women." *American Journal of Clinical Nutrition.* 1999;69:147-152.

Top Ten Nutritional Myths

The following study shows that elderly subjects who increase protein intake have a higher bone mineral density (BMD) measurement than those with a lower protein intake. Eating meat builds better bones.

"Calcium intake influences the association of protein intake with rates of bone loss in elderly men and women."
The American Journal of Clinical Nutrition.

> "Increasing protein intake may have a favorable effect on change in bone mineral density (BMD) in elderly subjects supplemented with calcium citrate malate and vitamin D."

Grilled Salmon and Green Beans

Chapter 3

Myth No. 8

Dietary Fat Should Be No More than 30% of Calories

This myth seems so true, innocent, wonderfully healthy, and wholesome that most people swallow it without question, but it is the most deadly myth of all dietary dogma and lies. The fat phobia is the result of worldwide brainwashing. When we reduce fats in the diet, our only option is to replace them with carbohydrates and begin the journey down the road to obesity, diabetes, heart disease, cancer, and the entire medical book of autoimmune diseases. Carbohydrates are slow to destroy the body, but they eventually prove to be extremely deadly. The time period between the start of a high-carbohydrate diet and disease onset has been labeled as the *20-year rule*.

> **Dr. David's low-fat diet has caused his hypoglycemia, heart disease, and degenerative disc disease.**

This is the true story of an acquaintance whose lifestyle has been according to the recommendations of the American Medical Association (AMA), American Heart Association (AHA), and the American Diabetes Association (ADA). But he still got hypoglycemia at age 53, plugged heart arteries at age 57, and degenerative disc disease at age 60. There are millions of similar cases each year in the United States, Canada, United Kingdom, and Australia.

Dr. David is employed at a hospital. His wife, Susie, was also well versed in standard medical recommendations. They have followed the AMA and AHA recommended low-fat diet religiously for most of their adult lives. They always shopped the low-fat selections in the grocery store and ate lots of "healthy" fruits and whole grains. They did recognize the necessity of eating meat but always selected the low-fat cuts, removed the visible fat on their plates, and always removed the skin from chicken.

Dr. David and Susie were "health nuts." Their busy schedule included jogging several times a week, but when it damaged Susie's knees, they switched to hiking and riding bikes on the lake road where they live. Since exercise was a major priority, they ran the local 10K race every year. When Susie could no longer run because of her knees, David continued to run the race religiously. It was the highlight of their healthy activity.

Dr. David and Susie were not fat or overweight so they never went on yo-yo diets. They never smoked because they knew it was unhealthy. Their bodies appeared to be in optimum health, but inside they were sick, sick, sick.

By age 53, Dr. David was having problems with blood sugar metabolism. His blood sugar surged when he ate his normal high-carbohydrate meals but plunged later and produced the typical symptoms found in pre-diabetes. Dr. David followed the recommendations of the American Diabetes Association (ADA) by snacking on granola bars and drinking fruit juice between meals in order to keep his blood sugar stable. Dr. David's biggest disadvantage was his medical degree because he believed the nonsense put forth by his professional societies. The low-fat diet is always high in carbohydrates and forces blood insulin levels to rise. Dr. David is suffering because he believed the "big fat lies" about the low-fat diet.

Dr. David was complacent about learning nutritional facts and didn't question the recommendations put forth by his professional societies. If he had done his own research into the

Chapter 3

history of heart disease, he would have discovered that because heart disease was so rare in 1900, it was not listed in the medical books as noted by the late cardiologist, Dr. Robert C. Atkins. Dr. David didn't realize that the cookbooks from that era show people eating lots of saturated fat in the form of pork lard, beef suet, butter, eggs, and coconut oil. Eating the skin from chicken and turkey was considered the best part of the bird. Dr. David still doesn't realize that the low-fat diet is the vegetarians' attempt to discourage the eating of animals and has nothing to do with healthy nutrition. He doesn't realize how deadly carbohydrates, such as fruit, bread, whole grains, cereals, sugar, flour, rice, legumes, potatoes, and soy really are.

Dr. David and Susie were proud that they considered themselves health and diet extremists. They followed the popular exercise and diet recommendations to the letter. They still think the *USDA Food Guide Pyramid* is a healthy way to eat because his professional medical societies fully support it.

Dr. David is now forced to decide which mechanical method he should use to unplug his heart arteries. He doesn't believe diet and exercise can help. What can he eat? He has been on the recommended low-fat diet for many years and his exercise has been unrelenting. His lifestyle was in perfect accordance with all the recommendations. His doctors will no doubt recommend the same diet for the **cure**, but it was actually the **cause** of his heart disease. Dr. David is in big trouble. What does a doctor do now? His erroneous training had double-crossed him and he doesn't even know it.

Low triglycerides, insulin, and LDL combined with High HDL cholesterol are vital for a healthy heart. This is only possible on the low-carbohydrate diet.

Top Ten Nutritional Myths

Dr. David's declining health hasn't stopped there. He developed degenerative disc disease in his back which required the insertion of rods between two vertebrae. He and Susie still eat lots of fresh fruit, dried fruit, and the yogurt that his profession recommends for healthy bones. Susie has been diagnosed with cancer and has symptoms of heart disease. They do not realize that their low-fat, high-carbohydrate diet is slowly killing them.

The Low-Fat Diet Causes Gallbladder Disease

> "The gallbladder resembles a pouch and collects green bile fluid as it drains from the liver through the common bile duct. This bile fluid is made, in part, to help with digestion (the salts in bile make it easier for you to digest fats). However, bile also contains some waste products including bilirubin (generated when old red blood cells are destroyed to make room for fresh ones) and cholesterol. Gallstones form when cholesterol or bilirubin particles begin to cluster together into a solid lump. The stone grows in size as the bile fluid washes over it, much like a pearl can form inside an oyster." Aetna InteliHealth.

The gallbladder collects bile fluid but does not completely discharge it when a low-fat diet is eaten. This bile fluid stagnates, and the gallstones formed from the precipitates become large enough to plug the discharge outlet of the bile duct. A single meal of fatty food can set off a gallbladder attack when it attempts to discharge its contents, including the stone or stones. If the pain becomes too severe, doctors will remove the entire gallbladder. However, the common bile duct at the junction of the liver bile duct and the gallbladder bile duct cannot be removed when it becomes plugged; therefore, the stone must be removed.

Chapter 3

A high-fat diet prevents gallstones because the gallbladder discharges the bile on a regular basis in order to digest fats. The system remains healthy on the high-fat diet. During the last 40 years, gallbladder disease has become epidemic as more people have switched to the low-fat diet. One woman on a nutritional message board stated, "Several friends and I decided to go on a low-fat diet. We didn't lose any weight, but we all lost our gallbladders."

Roast Chicken
Eat All of the Skin for the Essential Fatty Acids
Saturated Fats Inhibit Cancer Cell Growth

Top Ten Nutritional Myths

Myth No. 9

Soy Products Are Healthy Foods

The money and power behind the soy industry are monumental. Soy is heavily advertised; sales are growing at a tremendous rate, but soy has a sick nutritional foundation. Soy is not a healthy food. Soybeans in their natural state are not an edible food for animals, birds, or humans. The raw beans are highly processed in order to make the hundreds, if not thousands, of products based on a vegetable source of protein for the vegetarian market. However, this processing cannot remove all of the inherent unhealthy nutritional features of the products. Science is consistently finding serious problems with soy.

Soybean plants are the most genetically-modified organisms (GMO) in the United States' food supply. As of January 2006, the percentage of soybeans that has been genetically modified is estimated to be 75% and growing fast. Virtually all soy or soy protein products on supermarket shelves have been contaminated. Just because a soy product is labeled "organic" does not mean it is GMO free. Vegetarians gorge themselves on soy products; soy milk and soy burgers are among their favorites. Vegetarian foods, including breads, bagels, cereals, crackers, pastas, and energy drinks have all been fortified with soy protein made from GMO-contaminated soybeans.

The push behind the sale of soy products began when soybean oil replaced animal fats for cooking. However, soybean oil is an omega-6 fatty acid that has been scientifically proven to be pro-inflammatory and disease causing. (See Myth No. 10.) It is thought to be a major contributor to both heart and intestinal diseases. Soybean oil is widely used in fast-food restaurants for deep frying. After extracting the oil, manufacturers are left with

Chapter 3

soy meal, a waste product they want to sell. Soy meal is high in protein, and the push is on to sell this protein as a substitute for meat and fish. Soy products contribute to many health problems. They are full of isoflavones that function as strong estrogen agonists known as *phytoestrogens;* these disrupt normal sex hormone functions, cause hypothyroidism, and disrupt endocrine functions. Unfermented soy products contain enzyme inhibitors and unhealthy phytates that are not deactivated by cooking. Phytic acid combines with minerals in the intestines and prevents absorption.

Do not eat any soy products unless they are fermented or sprouted. Miso, tempeh, and natto are acceptable in very limited quantities. Better yet, don't eat any of these because you don't know if the required long fermentation process was used. It probably was not because our current manufacturing processes are so rapid. Do not use soy protein powders, but whey protein powders are highly recommended. Soy protein is missing several of the amino acids, one of which is classified as an essential. Do not substitute soy products for meat, fowl, fish, and seafood. Do not eat soy protein chips or cereals. Tofu made from soybeans has been shown to shrink the brain and cause cognitive impairment (brain fog).

> "The study, published in the latest issue of the *Journal of the American College of Nutrition*, showed that eating tofu two or more times a week affected brain function, leading to poor attention and memory as well as brain atrophy, or shrinkage." Health food can shrink brains of the middle-aged, says study. April 10, 2000, *The Independent*.

Top Ten Nutritional Myths

Myth No. 10

Omega-6 Vegetable, Seed, and Nut Oils Are Healthy

Professional nutritionists tell us to "Go nuts over nuts." The recommendation to eat large quantities of nuts is scientifically wrong. Absolutely avoid omega-6 polyunsaturated vegetable, seed, nut, and grain oils made from corn, soybean, safflower, sunflower, cottonseed, almond, apricot, grapeseed, hazelnut, peanut, poppy seed, rice bran, sesame, tea seed, tomato seed, walnut, and wheat germ. These are high-volume, high-profit oils. The companies involved make unrealistic and incorrect health claims. Contrary to these health claims, polyunsaturated omega-6 vegetable oils have been linked to breast and prostate cancer in clinical studies.

Omega-6 arachidonic fatty acid is an essential in the diet as the precursor of prostaglandins and leukotriene hormones; the amount obtained from meat, fish, and fowl far exceeds the requirement. Arachidonic fatty acid is the only one in the family of omega-6 oils that is essential in the human diet.

Lack of Omega-6 Arachidonic Fatty Acid Associated with Severe Dermatitis

> University of Illinois – April 14, 2010
>
> "A specific omega-6 fatty acid may be critical to maintaining skin health, according to University of Illinois scientists."
>
> "In experiments with mice, we knocked out a gene responsible for an enzyme that helps the body to make arachidonic acid. Without arachidonic acid, the mice developed severe ulcerative dermatitis. The animals were very itchy, they scratched themselves continuously, and

Chapter 3

> they developed a lot of bleeding sores," said Manabu Nakamura, a U of I associate professor of food science and human nutrition.
>
> When arachidonic acid was added to the animals' diet, the itching went away, he said."

Does your skin itch? Eat some high-fat meat. Pork shoulder roast is high in arachidonic acid.

The dietary goal must be to obtain adequate omega-3 fatty acids. Take one tablespoon of Carlson's® Lemon Flavored Cod Liver Oil per day as the best source for omega-3 fatty acids. It contains no omega-6 fatty acid as found in flax seed oil and is a great source for vitamin D.

Omega-6 fatty acids are pro-inflammatory, and they cancel the benefits of the good omega-3 fat. Flax is not the best omega-3 fat because our bodies must convert the shorter-fat ALA in flax to EPA and DHA before we can receive major benefits; some of us are unable to do this. Flax seed oil also contains a high level of omega-6 fatty acids which should be limited. Do not take flax seed oil or eat flax seeds.

> "A high omega-6/omega-3 ratio, as is found in today's Western diets, promotes the pathogenesis of many chronic diseases, including cardiovascular disease." "The omega-6/omega-3 fatty acid ratio, genetic variation, and cardiovascular disease." *Asia Pac J Clin Nutr.* 2008;17 Suppl 1:131-4.

You can certainly find studies, articles, and books that claim omega-6 vegetable oils are perfectly healthy. Some studies even claim omega-6 fatty acids are heart healthy. Many of these claims

Top Ten Nutritional Myths

are based on the assumption that omega-6 fatty acids displace saturated animal fats in the diet, and they say that saturated animal fats are unhealthy. The bottom line is pure conjecture and assumptions. Remember, the multi-billion dollar omega-6 vegetable oil industry can produce studies that claim anything they want.

Below is a list of the percentages of unhealthy polyunsaturated omega-6 fatty acids contained in the most commonly-consumed vegetable oils:

Safflower Oil	(80%)
Sunflower Oil	(68%)
Corn Oil	(57%)
Soybean Oil	(53%)
Cottonseed Oil	(53%)
Peanut Oil	(46%)

Omega-6 gamma-linolenic acid (GLA) is another bright spot in the omega-6 polyunsaturated fatty acid family. A person should supplement with omega-6 gamma-linolenic acid (GLA) by taking one capsule of borage oil a day with a meal. Borage oil contains a greater amount of GLA than primrose oil. The body does not produce essential fatty acids; they can only be ingested through food. GLA and omega-3 fatty acids together produce a type of prostaglandin, called *E1 series*. This type of prostaglandin helps reduce inflammation and aids in digestion. Solid scientific research shows omega-6 fatty acids are highly inflammatory and must be avoided by anyone with bowel disease, heart disease, arthritis, or any other autoimmune disease. Healthy people should seriously limit these omega-6 fatty acids. Dr. Robert C. Atkins' *Age-Defying Diet Revolution* and Dr. Michael Eades' *Protein Power Lifeplan* each contain long sections that describe the unhealthy effects of these oils.

Chapter 3

Bonus Myth No. 11

People Are All Different

This myth is so rampant that it just had to be added to the list of ten. The scientific design of the human body is held in very tight restraints. A special diet for a person with a different blood type or metabolic type is not effective because people are not different. Humans react in the very same way when they are given the same environment, the same original health condition, and the same nutrition. People only appear to be different because some have health conditions that restrict their normal body functions. A typical example is the diabetic who attempts to go on the low-carbohydrate diet or eats a large piece of high-sugar cherry pie with a glass of orange juice. His reaction will be much different from that of a healthy person. People think everyone is different because someone develops a disease and another family member does not. They don't realize that people in the same family can sit at the same table and eat completely different meals as defined by protein, fat, and carbohydrate content. A person can go into a restaurant and eat a low-carbohydrate dinner while others at the same table are eating a very high-carbohydrate meal.

Many people use a common cliché to sidestep scientific absolutes by saying, "Oh, that may work for them or for you, but that doesn't work for me at all." This is the typical logic people use when they fail to realize that the case in point violates universal laws of science. Instead of challenging or immediately dismissing this excuse, nearly everyone readily accepts it. The myths and nonsense about nutrition and health go on and on.

People search for excuses to continue their high-carbohydrate addictions. One of the most popular excuses they give in order to

Top Ten Nutritional Myths

continue their addictive lifestyles is to simply say, "All people are different." These people are deceiving themselves.

People seem to go ballistic when told they are the same as everyone else. The excuse for their obesity or diseases is snatched away by the scientific truth that everyone is basically the same. They know they must hang onto the excuse that they are somehow different. They want a scapegoat for their problems so they maintain, "It's not my fault. People are different." They want to blame their conditions on anything but their diet. When every other excuse has been exhausted, they resort to the nonsense that can't be readily disproved, "It is in my genes."

Healthy High-Protein and High-Fat Foods

All Fresh Meat	Beef	Lamb
Pork	Chicken	Turkey
Wild Game	Fish	Seafood
Shell Fish	Hard Cheeses	Coconut Oil
Butter	Organ Meats	Eggs

Health experts tell us to eat a diet with less protein and more carbohydrates based on our blood or metabolic type. These claims are scientifically false. Studies of primitive people show that the high-protein, low-carbohydrate diet is healthy for every individual but the high-carbohydrate diet is unhealthy for everyone. The Egyptian mummies have provided us with solid scientific evidence that a high-carbohydrate diet with an abundance of whole grains, fruits, and vegetables results in many debilitating diseases and death at an early age. My diet program is ideal for everyone of any blood type, metabolic type, or age.

Chapter 3

Yes, some differences are in the genes, but blood type or metabolic type do not make eating carbohydrates and fiber healthy for anyone.

Absolute Truth Exposed

Volume 1

Chapter 4

Top Twelve Historical Events That Created Our Current Health and Nutritional Quagmire

Chapter 4

These top twelve historical events that created our current health and nutritional quagmire may have occurred with the good intentions of everyone involved. They simply could not have known the unintended consequences. Others may have had evil intentions behind their creation, including profit, political, or religious motives without regard to the consequences. Their primary goals were much more important to them than the health of those who would suffer from these consequences. This atmosphere continues to this day as the driving force in our medical and nutritional fields. Coming forth with solid scientific, medical, and nutritional truths that don't agree with the politically-correct nutritional dogma has been met with a barrage of criticism and threats as experienced by Dr. Robert C. Atkins for thirty years and William Banting from London, England, in the mid nineteenth century.

Nutritionists are consistently being guided and controlled by the profit or religious motives of special interest groups. This unscientific approach to nutrition has led to many conflicting conclusions about optimal human nutrition. Nutritional research is commonly carried out in a very sloppy manner with the use of programs designed to give the researcher the biased result he is seeking. For this reason, nutritional research arrives at opposite results and lacks repeatability. Lies, distortions, and fraud are rampant in these studies, and the result is a serious loss of confidence in the conclusions reached by any nutritional study. Diet and nutritional studies are great examples of garbage in— garbage out.

Draining a patient's blood as a method of healing is on the list of medical theories and practices that were later proved to be false. This past practice was later proved to be very detrimental to the patient, but it caused a huge—although unknown—number of

Top Twelve Historical Events

deaths. Patients had their blood drawn until they literally bled to death. The exact cause of US President George Washington's death in December 1799 has been hotly debated. He contracted a cold or pneumonia after he was caught in a winter storm while riding his horse at his estate in Mt. Vernon, Virginia. (He acquired this estate upon his marriage to Martha Dandridge Curtis, a rich widow. She was 27 years of age, and George was 26 at the time of their marriage.) A total of 80 ounces of blood was drained from the father of our country over a 12-hour period according to David Morens, MD, a National Institutes of Health epidemiologist who has written articles on the President's demise. The doctors hastened, contributed to, or caused his death by the bleeding which most likely led to shock.

At the present time, the American Medical Association (AMA) and the majority of physicians believe that eating fats—especially saturated animal fats—produces heart disease, cancer, diabetes, osteoporosis, and intestinal diseases. This theory is upside-down and backward as was the past medical practice of draining the patient's blood to promote healing. The AMA refuses to come to terms with the scientific fact that these diseases are caused by a high-carbohydrate diet deficient in protein and fats from animal products. Deaths caused by the errors of the medical profession and prescription drugs continue at an alarming rate and may actually be increasing just as in President George Washington's day. Deaths are just the tip of the iceberg when you compare them to health problems that leave patients just short of death as a result of errors and drugs.

Dr. Joel D. Wallach, DVM, ND, states in *Rare Earths - Forbidden Cures* that the death rate in Los Angeles, California, dropped approximately 30% in 1976 during a seven-week doctors' strike.

Chapter 4

Two weeks after the strike ended, the death rate doubled from the strike's low. (See page 237 in his book.)

The Industrial Revolution is the master (or macro) historical event that precipitated many other highly unhealthy events which transformed natural, healthy food into disease-causing products. When unsuspecting people consume these products, they are totally unaware of their long-term effects.

The following historical events that destroyed our food supply are listed chronologically, not by the magnitude of the harm inflicted on our health.

Modern Fruit Juice Packaging Plant

Top Twelve Historical Events

Event No. 1

The Industrial Revolution

The Industrial Revolution is a double-edged sword. Modern machines can keep fresh meat, fowl, fish, seafood, fruit, and vegetables refrigerated or frozen for safe storage until ready for consumption. These foods are available from distant lands in great variety and abundance. Prior to the Industrial Revolution, animals were kept alive until they could be slaughtered and quickly sold to the consumer. Great cattle drives north from Texas to Chicago during the nineteenth century are a good example. In many countries today, animals are walked to the marketplace where they are slaughtered and quickly sold without refrigeration.

The downside of the Industrial Revolution has been the creation of "food" that is the output from a chemical plant. These foods are generally made from carbohydrate-laden grains, but some are made from the vegetable protein and vegetable fats. Most of the people from past centuries would not consider eating the products in our modern supermarkets.

Another downside of the Industrial Revolution is the year-round availability of high-carbohydrate fruit, which is highly perishable. Fruit was consumed in its season prior to the Industrial Revolution and did not present the health hazard we see today. Canning fruit increases the unhealthy effects because refined sugar is generally added which results in a can or bottle of "food" that has little nutritional value. It is basically all sugar in the form of sucrose and fructose.

The following study is repeated from the previous chapter due to the constant onslaught of claims that fruit is a very healthy food. It is not. Fructose in fruit causes insulin resistance (metabolic

Chapter 4

syndrome), diabetes, heart disease, cancer, inflammatory bowel disease, and Alzheimer's disease.

"Hypothesis: fructose-induced hyperuricemia as a causal mechanism for the epidemic of the metabolic syndrome."

"The increasing incidence of obesity and the metabolic syndrome over the past two decades has coincided with a marked increase in total fructose intake. Fructose--unlike other sugars--causes serum uric acid levels to rise rapidly. We recently reported that uric acid reduces levels of endothelial nitric oxide (NO), a key mediator of insulin action. NO increases blood flow to skeletal muscle and enhances glucose uptake. Animals deficient in endothelial NO develop insulin resistance and other features of the metabolic syndrome. As such, we propose that the epidemic of the metabolic syndrome is due in part to fructose-induced hyperuricemia that reduces endothelial NO levels and induces insulin resistance. Consistent with this hypothesis is the observation that changes in mean uric acid levels correlate with the increasing prevalence of metabolic syndrome in the US and developing countries. In addition, we observed that a serum uric acid level above 5.5 mg/dl independently predicted the development of hyperinsulinemia at both 6 and 12 months in nondiabetic patients with first-time myocardial infarction. Fructose-induced hyperuricemia results in endothelial dysfunction and insulin resistance, and might be a novel causal mechanism of the metabolic syndrome. Studies in humans should be performed to address whether lowering uric acid levels will help to prevent this condition."

Top Twelve Historical Events

Event No. 2

The Flour Mill

The grain grinding wheel was used to make flour for bread many centuries before the Industrial Revolution. In Biblical history, Samson was captured by the Philistines in the eleventh century B.C. after his lover, Delilah, cut off his hair that gave him his miraculous strength. He was bound into service and put in prison where he turned a grinding mill. At that time, the mills were powered by animals, prisoners, or slaves but were later powered by water in the Roman Empire and by wind in Holland. The Industrial Revolution improved the design of water-wheel-powered grinding mills, and the steam engine was adapted to turn them. The advancements made the grinding mill a very common piece of machinery which could be set up anywhere—thus increasing the consumption of grains in the diet.

Carbohydrates Are Extremely Addictive

Chapter 4

Old Waterwheel-Powered Grain Mill

Modern Grain Mill

Top Twelve Historical Events

Grains have always been a problematic food source for humans. The ancient Egyptians had a high level of grains in their diet and suffered all of the same diseases we see in our modern high-grain diets. They even describe the symptoms of diabetes and heart disease in their writings. Their mummified bodies which are available to us for medical examination exhibit serious health problems, including osteoporosis (degeneration of the spine), rheumatoid arthritis, and highly decayed teeth.

In the Bible, Joseph, the son of Jacob, was sold into slavery in Egypt by his brothers. The Pharaoh recognized Joseph's management abilities and placed the entire economy into his hands. Joseph ordered large quantities of grains to be stored as a hedge against an impending drought which later came upon the entire region just as he had expected. This story illustrates the benefit of grain storage for an emergency food supply. The story does not imply that the grains provided optimal health because the Egyptian mummies are evidence that it does not. The Nile River Delta provided an exceedingly abundant supply of grains that contributed to the poor health of the Egyptian people.

Grains cause a large number of human health problems, and many people are unaware that their inability to tolerate gluten is affecting their health. Bakeries use high-gluten grains because they make the bread rise faster and higher and produce a better appearance and a softer texture. The proteins in grains set off autoimmune reactions and disease when they cross from the small intestine or colon into the bloodstream in a condition known as *leaky gut syndrome*. The grains feed intestinal bacteria and yeasts and promote the proliferation of pathogenic bacteria in the colon. Leaky gut syndrome triggers a vicious cycle of many serious diseases. Grains are not the wonderful food that manufacturers would like us to believe. Grains do not contain

Chapter 4

enough fats and protein to prevent deficiencies, and the high levels of carbohydrates provide empty calories which make people obese and unhealthy. Whole and refined grains make delicious breads, pastries, and desserts but they promote disease. Whole grains are not healthy as commonly believed.

Grain milling and processing have led to greatly increased carbohydrate consumption in all English-speaking countries. Grains are among the major factors in the epidemics of obesity, diabetes, heart disease, cancer, intestinal diseases, osteoporosis, and numerous autoimmune diseases. Today's grain processing plants are not simply grain mills; they are complex chemical plants that produce hundreds of products from grains.

Modern Highly-Productive Wheat Farm

Top Twelve Historical Events

Event No. 3

Refined Sucrose Sugar

Immediately after its discovery, sucrose sugar became an instant success and a highly-desired sweetener. It was first considered to be a spice that tasted like honey from plants instead of bees. The sucrose molecule is comprised of one glucose sugar molecule and one fructose sugar molecule.

"The History of Sugar"
http://mimi.essortment.com/historyofsugar_rzow.htm

> "The discovery of sugarcane, from which sugar, as it is known today, is derived, dates back unknown thousands of years. It is thought to have originated in New Guinea, and was spread along routes to Southeast Asia and India. The process known for creating sugar, by pressing out the juice and then boiling it into crystals, was developed in India around 500 BC."

> "Its cultivation was not introduced into Europe until the middle-ages, when it was brought to Spain by Arabs. Columbus took the plant, dearly held, to the West Indies, where it began to thrive in a most favorable climate."

> "It was not until the eighteenth century that sugarcane cultivation was began in the United States, where it was planted in the southern climate of New Orleans. The very first refinery was built in New York City around 1690; the industry was established by the 1830s. Earlier attempts to create a successful industry in the U.S. did not fare well; from the late 1830s, when the first factory was built, until 1872, sugar factories closed down almost as quickly as they had opened. It was 1872 before a factory, built in

Chapter 4

California, was finally able to successfully produce sugar in a profitable manner. At the end of that century, more than thirty factories were in operation in the U.S."

Sugar Cane Farm When Plants Are Young

However, sugar is not the sweet, wholesome food it appears to be. It is one of the most unhealthy and highly addictive foods one could possibly eat. Eating sugar is unhealthier than smoking cigarettes. All societies that ingested high quantities of sugar suffered as a result. It is possible—although not likely because of the sugar lobby—that court rulings of the future will classify sugar as the root cause for many diseases and will impose restrictions and assess punitive damages. Sugar is following in the footsteps of cigarettes.

Fruit and fruit juices are not much healthier than refined sugars. Yes, they have vitamins and minerals, but these can easily be obtained in the diet from low-carbohydrate foods, such as non-starchy vegetables, meat, fish, and fowl. Orange, cherry, and grape juices contain a high concentration of fructose, and the body's reaction to it is identical to its reaction to high-fructose corn syrup sweetener—insulin resistance! Many of the commercial juices are watered down, and high-fructose corn syrup and sucrose sugar are added. These could best be described as fruit-flavored sugar water drinks. The gullible consumer has been brainwashed into believing the fructose in fruit is healthier than fructose in sugar. Drinking 100% fruit juice does not make it healthier.

Fructose has been proven to cause insulin resistance, the precursor of Type 2 diabetes. Researchers have known for decades that feeding rodents high-fructose corn syrup will create insulin resistance. This procedure is used to prepare the animals for accelerated diabetes research.

Sugar Cane Field at Harvest Time

Chapter 4

Event No. 4

Dr. Ancel Benjamin Keys and the Low-Fat Diet/Heart Disease Hypothesis

Dr. Ancel Benjamin Keys (January 26, 1904 – November 20, 2004) was a University of Minnesota physiologist who was very influential in promoting the false idea that dietary saturated animal fats cause heart disease. He conducted experiments and launched his first attack against saturated fatty acids in *Biology of Human Starvation*. He promoted polyunsaturated fatty acids as found in vegetable oils. More recent studies have shown vegetable oils to be unhealthy. He also effectively promoted the false theory first proposed by British surgeon, Dr. Dennis Burkitt, that the high-fiber diet was healthy. Dr. Ancel Keys' unsubstantiated dietary theories have probably resulted in more deaths worldwide than Russia's Joseph Stalin and Germany's Adolph Hitler combined.

Dr. Ancel Keys was a political dietitian. He skipped over the true science and faked the results from his "scientific studies." He was a bought man as his funding and grants came from the sugar and cereal industries. (Sugar and cereal are low fat products; cereal is a high-fiber product.) He effectively changed the diet of the world—especially in all English-speaking countries. The death rates from heart disease and cancer have soared since Dr. Ancel Keys successfully brainwashed the world with his unsubstantiated dietary ideas. Obesity and diabetes have become epidemic on his low-fat, high-fiber diet with no end in sight to the upward-spiraling statistics. Obesity and heart disease were both quite rare before the world heard of Dr. Ancel Keys. His diet recommendations also led to many autoimmune diseases, which typically result in years of suffering for people who have

Top Twelve Historical Events

rheumatoid arthritis, asthma, Crohn's disease, ulcerative colitis, multiple sclerosis, psoriasis, lupus, and many others.

Ironically, Dr. Keys lived to 100 years of age. Many would argue that his longevity is proof that his anti-saturated fat campaign was correct. This is a very weak argument because he may not have eaten according to what he preached. The life span of one person does not make a controlled study, and controlled study results are often fraudulent. Millions of people who followed his dietary advice died at an early age.

Healthy Fatty Acid Foods

Animal Fats	Arachidonic Acid	Saturated Fats
Fish Oils	EPA Omega-3	DHA Omega-3
Butter	Coconut Oil	Monounsaturated

Unhealthy Fatty Acid Foods

Grain Oils	Seed Oils	Most Nut Oils
Vegetable Oils	Omega-6 Fats	Polyunsaturated
Trans Fatty Acids	Soybean Oil	Corn Oil

Low-Fat Keys

Chapter 4

Event No. 5

Ball® Canning Jars and Fresh Fruit

In 1880, the five Ball brothers founded the Wooden Jacket Can Company®, which manufactured wooden-jacketed tin cans for paints and other products. In 1884, the renamed Ball Brothers Glass Manufacturing Company® began making yet another new product—the glass-lined tin home canning jar that would one day make Ball a household name. Ball acquired the first of several small glass companies in 1898 and began making glass-blown canning jars. In 1909, they printed the first *Ball Blue Book*, which featured home canning recipes and techniques.

Ball® jars became highly popular for home canning jams, jellies, and fruit. The canning process was very easy and cost only a few cents per quart. Fruits such as peaches or apricots were blanched in hot water, peeled, cut in half, pitted, and placed in a quart jars. About 1-1/2 cups of syrup, containing two parts water and one part refined white sugar, were added to each jar. New seal tops were placed on the jars and secured by tightening the screw-on,

Top Twelve Historical Events

reusable top rings. The jars were then placed in a hot-water canning steamer or pressure cooker. A slightly domed lid indicated that the jars had lost their vacuum during storage and that the contents should not be eaten. Jams and jellies were made by heating a ratio of 2 cups of fruit with 1 1/2 cups of sugar. The mixture was smashed or juiced as desired, placed in the Ball jar, and put through the canning process. The sugar content in the preserved product was very high.

The resulting preserved fruit had very little food value and was essentially composed of sugars from the fructose and sucrose. The consumption of sugar skyrocketed as a result of the invention of home canning jars and paralleled the sharp rise in diseases, such as diabetes, cancer, heart, and intestinal diseases. The obesity epidemic had begun.

Canned fruit was also used as the filling for homemade pie. The crust was made with refined white flour, refined white sugar, and shortening (a combination of vegetable oils and trans fat hydrogenated vegetable oils). Crisco® was the most popular brand for these factory-made chemical fats. The resulting dessert was very delicious to the taste, but the grains, sugars, and trans fats together formed the "Deadly Food Pie."

Many diet gurus worship fresh fruit as the most wholesome food one can eat, and they often recommended it more strongly than whole grains. High-carbohydrate diet gurus and official US government information sources never criticize fruit. We are constantly hearing recommendations to eat lots of fruits and vegetables to achieve optimum health. Reports are regularly being released which quote new studies that say fruit prevents many illnesses. They claim that orange juice reduces elevated blood pressure and other fruits reduce the risk for cancer or heart disease. All of these claims are **myths, distortions, and**

Chapter 4

lies. Fruit is basically ***Nature's Candy***. Fruit is concentrated fructose sugar. It contains only carbohydrates and no proteins or fats. Addiction to high-sugar fruit is a serious problem among adults and children. A study of the shopping carts of adults and the eating habits of children easily proves that this unhealthy addiction exists in all age groups.

Deadly Food Pie
Sugar – Carbohydrates – Fiber – Vegetable Oils

The few healthy vitamins, minerals, and anti-oxidants found in fruit are praised to the heavens. Some fruit worshippers (called *fruitarians*) have attempted to live on a fruit-only diet but immediately found that their health crashes. Fruit has no essential amino acids and no essential fats necessary for life.

The main problem with fruit is the large quantity that many people eat. A little fruit goes a long way and should be severely restricted for a healthy diet. Having fruit available in the

Top Twelve Historical Events

refrigerator and supermarket 24 hours a day for 365 days a year is strictly contrary to the diet on which mankind has existed for many thousands of years. Fruit is now delivered to the supermarket from all continents of the Earth. Only 100 years ago, fruit was strictly a seasonal food at best and not available at all in many civilizations where they thrived in excellent health. Fruit does not grow in northern climates, at higher altitudes, or in many arid regions of the Earth. Years ago, much of the fruit was different from the hybrid varieties available today. Most of the natural fruit of the past was low in fructose and sour to the taste. Sweet fruits like grapes were turned into wine because all fruit is extremely perishable. Today's hybrid fruit has been created in the laboratory to have the highest sugar content possible in order to satisfy the addictive cravings of the general public. The trend continues to create an ever-increasing sweetness by genetically modifying the hybrid varieties.

Laboratory experiments have proved that fructose produces insulin resistance, which leads to diabetes and higher rates of heart disease, cancer, and intestinal diseases. Fruit is a very unhealthy food. In *Dr. Bernstein's Diabetes Solution,* Dr. Richard K. Bernstein says that he has not eaten fruit in 25 years, and this has made his health better.

Chapter 4

Event No. 6

Sugar-Sweetened Soft Drinks

The carbonated sugar drink we now call a *soda* or *soft drink* originated in 1886 as a soda fountain beverage called *Coca-Cola®*. The drink sold for five cents a glass, which was a high price at the time and equal to nearly one hour's wage for an unskilled laborer. The carbonation has no positive or negative health effect, but the high level of sugar is a serious health problem. Coca-Cola® was a huge success, and the sweetener (refined sugar) had become very plentiful at a low price. People quickly became "hooked" on the refreshing drink. Carbohydrates are highly addictive, and many people have lived their entire lives with an insulin rush from excessive glucose intake. Insulin is the body's most powerful hormone and makes resisting carbohydrates extremely difficult. Many would rather be sick than give up their sugar soft drinks and fruit with the high levels of fructose—the second most addictive carbohydrate. These addictions must be broken if we are to be healthy.

Other foods also contained the new refined sugar, but since Coca-Cola® was available throughout the day, it was consumed between meals. Most other high-sugar foods were served as desserts, and their consumption was limited.

The second major development for cola soft drinks occurred when sucrose sugar was replaced with high-fructose corn syrup. Sucrose sugar is half glucose and half fructose, about the same as high-fructose corn syrup. Diet carbonated soft drinks were a major improvement because sucrose and fructose were eliminated, but the artificial sweeteners that replaced these sugars contain ingredients that also cause health problems.

Top Twelve Historical Events

Event No. 7

The Steel Reinforced Concrete Grain Storage Silo

The first steel reinforced concrete grain storage silo was installed in 1900 and was an immediate success. Grain could be stored throughout the winter where it was kept dry and protected from rodents. During the off-season, it was reclaimed in perfect condition. These grain storage silos were installed on the great plains of the United States close to the harvesting areas. They now appear as clusters of huge concrete structures towering on the horizon.

The grain storage silo reduced the overall cost of grain products and increased their consumption. However, the health of a society declines in direct proportion to the amount of grain consumed by its people; the more grain produced, the more damage is done to the health of the people. Grain production has now become completely mechanized, from planting the seed on the farm to placing breads and cereals in the packages for sale.

Grain Storage Silos Nicknamed "Grain Elevators"

Chapter 4

Event No. 8

Vegetarians Sylvester Graham, Mrs. Ellen G. White, and the Kellogg Brothers

Sylvester Graham was a religious vegetarian who didn't have a clue about human nutritional requirements—yet he gathered thousands of zealous followers. He invented Graham® flour and the Graham Cracker® in 1829, and children switched from a healthy diet of animal protein and animal fats to one of unhealthy carbohydrates. Modern-day Graham® crackers would make Mr. Graham turn over in his grave. Although his cracker was very unhealthy, it did not contain the high level of sugar and deadly hydrogenated fats that are found in many of today's cracker products. Cracker manufacturers have begun to remove hydrogenated fats due to the public outrage against these chemical fats.

Mr. Graham's erroneous teaching influenced other vegetarians like the Kellogg brothers to follow in his footsteps. Graham preached against sexual desire. He referred to a sex drive as "venereal excess" or "aching sensibility," and claimed his Graham products would cure these "dire physiological" problems. According to the rumors, his sexual desire declined to zero, and in 1851 he died at the age of 58. His death was most likely caused by his faulty diet.

Mrs. Ellen G. White was the founder of the Seventh-Day Adventist Church®. In 1864, she announced that she had a spiritual vision against eating meat and butter. She was accused of copying the vegetarian diet philosophy of Dr. J. C. Jackson, but she denied this in church papers. However, her denial does not appear to be credible. She obviously got her vegetarian diet plan from Dr. Jackson's book which she had in the house, not from

Top Twelve Historical Events

Holy revelation. The vegetarian diet is a violation of the Holy Scriptures of the Bible. Dr. Jackson also invented the first cereal made from grains, which he called *granula,* as a substitute for meat and eggs in the diet. Many people give the Kellogg brothers credit for inventing cereal, but their discovery was making the cereal into flakes.

"Mrs. White's Health Visions: Was it God? Or Dr. Jackson?"
http://ellenwhite.org/refute9.htm

> "I did not know that such a paper existed as The Laws of Life, published at Dansville, N.Y. I had not heard of the several works upon health, written by Dr. J. C. Jackson, and other publications at Dansville, at the time I had the view named above. I did not know that such works existed until September, 1863, when in Boston, Mass., my husband saw them advertised in a periodical called the Voice of the Prophets, published by Eld. J. V. Himes. My husband ordered the works from Dansville and received them at Topsham, Maine. His business gave him no time to peruse them, and as I determined not to read them until I had written out my views, the books remained in their wrappers."

In 1869, Mrs. White made a public pledge to refrain from eating meat, but years later she personally wrote many times about eating deer, duck, oysters, herrings, chickens, and eggs. This was many years after her vision and recommendation to her followers to avoid such foods. In 1894, Mrs. White again made another public pledge against eating animal products. She died in 1915.

Chapter 4

"Oysters and Herrings - Did Mrs. White Practice what she Preached on Diet?"
http://www.ellenwhite.org/contra6.htm

> "I suppose you will be interested to know how we spent Christmas. Christmas morning we all took breakfast together--James Cornell; Florence and Clara, their two girls; Brother and Sister Moore and their three children; Sister Bahler and Etta, a girl living with them; and Sister Daniells, our cook, Father, and myself. We had a quarter of venison cooked, and stuffing. It was as tender as a chicken. We all enjoyed it very much. There is plenty of venison in market." (Manuscript Releases Vol. 14, p. 318-319, written December 26, 1878, from Denison, Texas, to "Dear Family at Battle Creek--Willie, Mary, Aunt Mary, Edith, Addie and May, and Brother and Sister Sawyer.")

Cheating on the vegetarian diet is more common than unusual. Vegetarians typically "try to cut back on meat" while they suffer from a guilt complex for eating it. Apparently, the guilt is sufficient suffering for them to continue to indulge in a food they falsely swear is unhealthy and turns putrid in the body.

John Harvey Kellogg was a young man in Mrs. White's Church, and he received financial support from her for his medical education. He became a doctor and manager of the Battle Creek Sanitarium in Michigan, USA. As a world-renowned physician, surgeon, and health reformer, he invented *protose* (a meat substitute that is currently made from nuts, beans, cornstarch, and seasonings) as well as other meat substitutes. In addition, he was a prolific writer and published more than fifty books.

Top Twelve Historical Events

"Kellogg's: How It All Began"
http://press.kelloggs.com/index.php/company_info/beginning.html

"Shortly after the turn of the century Dr. Kellogg came into conflict with the leaders of the General Conference over his attempt to get the control of all SDA medical institutions with which he had been associated. He finally did succeed in getting control of the Battle Creek Sanitarium, the Battle Creek Food Company, and the health institution in Mexico. He also began teaching strange doctrines. His book The Living Temple was permeated with the principles of pantheism. Everything was done to help him see his error. Ellen G. White worked with him personally and sent him many messages, but in vain. In 1907 he lost his membership in the church. Only a few intimate friends followed him." (SDA Bible Commentary, Vol. 10, art. Kellogg, John Harvey, p. 723. Emphasis supplied.) Pantheism and the "Alpha of Apostasy."
http://ellenwhite.org/egw67.htm

"William Kellogg's accidental discovery of cereal in 1894 marked the beginning of a multi-national food empire. While experimenting with different food production techniques at the Battle Creek Sanitarium in Michigan, William and his brother, Dr. Kellogg, decided to run boiled wheat dough through rollers which enabled them to produce thin sheets of wheat. After a sudden interruption in their laboratory activities left cooked wheat exposed to the air for more than a day, the Kellogg brothers decided to run the wheat through the rollers despite the fact it was no longer fresh. To their amazement, instead of a

Chapter 4

single, large sheet of wheat, the rollers discharged a single flake for each wheat berry—cereal flakes were born."

"Prompted by William Kellogg's belief that diet played an important role in a healthy lifestyle and that breakfast is the most important meal of the day, he spent the next eight years further developing the newly invented cereal, culminating in its first production in 1906. That same year, in order to increase awareness of the revolutionary product, W. K. bought a full-page ad in the July issue of the Ladies' Home Journal. As a result, sales leapt from 33 cases to 2,900 cases per day."

The Kellogg brothers both lived beyond 90 years of age. One would hesitate to criticize their diet recommendations because of their longevity, but an acquaintance who is a vegetarian in the SDA church was forced to apply for full disability at age 48 as a result of degenerative disc disease. He may live to age 90 as well, but his handicap and constant pain will make his life miserable for 40 more years. The Kellogg brothers may have lived a "clean" vegetarian lifestyle, but maybe they did not. People do not always follow their own advice.

Modern-day grain cereals made by Kellogg's® and other companies contain a very high percentage of sugar. The formulas of some cereal manufacturers list sugar as the leading ingredient on the package. Other blends may list a grain first but also list several different sugars, such as sucrose, fructose, and high-fructose corn syrup. The combination of all the different types of sugars could easily become the major ingredient. These products could best be described as grain-flavored sugar. They sometimes contain hydrogenated vegetable or seed oil (trans fats).

Top Twelve Historical Events

Had it not been for Dr. Kellogg, Mrs. White and her Seventh-Day Adventist Church® would have had little effect on the nutrition and health of a nation because the organization was small. He went on to found the giant Kellogg's® food manufacturing company that specializes in grain-based breakfast cereals. His efforts helped the entire English-speaking world to switch from meat and eggs to grain cereals for breakfast. The meat and egg breakfast was composed of protein and fat with no carbohydrates, while the grain cereal breakfast was composed of nearly all carbohydrates with very little protein and fat. The heart disease rate was close to zero, and virtually no medical books or journals mentioned anyone having a heart attack prior to the advent of grain cereals in 1900. The heart attack rate soared after the introduction of grain cereals and the refined sugar commonly used to sweeten them. The switch to the high-carbohydrate diet was underway. Dr. Kellogg was certainly a health reformer. His diet philosophies may have contributed to early diseases and early deaths of millions of people.

Breakfast grain cereals cost approximately $4.00 per pound at supermarkets in the USA, but the grain used to make the cereal only costs $5.00 per bushel @ 60 pounds per bushel or seven cents per pound. The raw grain is only 2% of the cost of cereal; the other 98% is manufacturing cost, packaging cost, distribution cost, and profit. Because the cost of the package is undoubtedly much more than the cost of the grain used to make the cereal, breakfast cereals are heavily advertised on children's TV programs. Children whine and cry in the supermarket when they see the box with the cartoon characters they remember from the clever TV advertisements, and Mommy simply throws it in the shopping cart to keep them quiet. The result is our current childhood obesity epidemic.

Chapter 4

Event No. 9

Hydrogenation of Vegetable and Seed Oils

Vegetable and seed oils are primarily comprised of polyunsaturated and monounsaturated fatty acids. These oils are liquid at room temperature and can easily become rancid because of their non saturation. The vacant bonds in their makeup of carbon and hydrogen atoms allow undesirable bonding sites for other atoms or molecules. Unsaturated fats become rancid and unhealthy when oxygen or a free radical molecule bonds to the vacant site. Saturated fats have all of the carbon sites bonded to a hydrogen atom, which makes the fat inert against other atoms or molecules. This gives the saturated fatty acids chemical stability and also makes them solid at room temperature. The saturation makes them healthier for us to eat.

In 1911, it was discovered that hydrogen atoms could be forced to attach to the unsaturated vegetable and seed oils by applying pressure and heat in a vessel in which hydrogen was percolated through the oil. The result was not a true saturated fat but a product called *trans fat*, which has many of the characteristics of saturated natural fat but is a foreign molecule to the body. Trans fats are chemical lipids rarely found in nature. This process produced a product that could be substituted for saturated fat in food at a much lower cost. Greedy food producers began using trans fats in food products without regard to their health effects. We now know that these unhealthy factory-made chemical fats damage the cells of the body and produce many diseases.

Trans fat manufacturers attacked natural animal fats with lies and distortions in order to promote their chemical fats. As usual, the gullible public bought the lies hook, line, and cracker. Animal rights advocates and vegetarians jumped on the bandwagon because they promote any lie or distortion ever conceived in their attempts to slow the general consumption of animal products. The lies are almost limitless.

Linoleic fatty acid is a precursor of arachidonic acid (an omega-6 fatty acid) which is a precursor of the prostaglandin E2 series (PGE2). Other omega-6 fatty acids are also precursors of PGE2 that reduces the ability of macrophages and natural killer (NK) cells to kill cancer cells and tumors. Since omega-6 fatty acids are found in abundance in grains, nuts, and seeds, these foods should be strictly limited. Hazel nuts, pine nuts, and macadamia nuts are healthier because they are low in omega-6 fatty acids. However, arachidonic acid is essential in the diet and has many health benefits.

The edible oil industry produces omega-6 polyunsaturated vegetable, seed, and grain oils, such as corn, soybean, safflower, sunflower, peanut, and cottonseed oils. It is highly suspected that these oils are among the leading causes of heart disease and cancer—both of which increase in concert with increases of omega-6 fatty acids in the diet. Omega-6 fatty acids are pro-inflammatory and proven to cause or contribute to a long list of autoimmune diseases, such as asthma, rheumatoid arthritis, lupus, multiple sclerosis, Crohn's disease, fibromyalgia, irritable bowel syndrome, inflammatory bowel disease, and many others.

"Postmenopausal breast cancer is associated with high intakes of omega-6 fatty acids (Sweden)"
http://www.ncbi.nlm.nih.gov/entrez/query.fcgi?cmd=Retrieve&db=PubMed&list_uids=12588084&dopt=Abstract

> "RESULTS: Saturated fat and the omega3-omega6 fatty acid ratio were not related to increased risks, but positive trends were seen for total ($p = 0.031$), monounsaturated

Chapter 4

(p = 0.002), and polyunsaturated fat (p = 0.0009), especially omega6 fatty acids and the polyunsaturated-saturated fat ratio (p = 0.004). With mutual adjustment for different types of fat, an elevated risk remained significant in the highest omega6 fatty acid quintile (RR= 2.08, 95% CI 1.08-4.01)."

The above study proves that saturated fats and omega-3 fats as found in red meat and fish were not associated with an elevated risk for breast cancer in postmenopausal women; but monounsaturated omega-9 fats as found in olive oil, and polyunsaturated omega-6 fats as found in grains, seeds, and nuts increased the risk of breast cancer—specifically omega-6 fats. These excellent results are directly opposite to the myths, distortions, and lies promoted by vegetarians and manufacturers of high-carbohydrate foods. Beef, lamb, and pork with saturated fats are very healthy foods.

Bottle of Yellow Safflower Oil

Top Twelve Historical Events

Event No. 10

The Grain Drill and Combine Harvester

Grain Combine Harvester

One normally does not think of the grain drill—a machine used for planting grains—or the combine harvesting and threshing machine as having any detrimental impact on human health. This would be a correct assumption if grains were a healthy farm product, but they are not. The case against grains and refined grain products is gaining momentum on a daily basis. The grain drill and combine harvester are closely tied to the grain flour mill. These machines created a surge in the consumption of grain products and greatly reduced their prices. The increased production of grain products leads to the displacement of healthy foods, such as meat, fish, fowl, and seafood. Buyers were more price conscious than health conscious until the onset of our obesity and diabetes epidemics. Shoppers are now more health conscious but continue to be hammered with a constant barrage

Chapter 4

of false health and nutritional information by special interest groups who deceive them by claiming grain products are healthy.

People are turning to "healthy" cereal that contains nuts and dried fruit but does not contain sugar. They are being tricked again. Grain cereals have the same amount of carbohydrate calories per pound as sugar. The mantra "good carbs – bad carbs" is a false chant. Some carbohydrates are bad and the rest dreadful.

On July 15, 1929, Toronto inventor, Thomas Carroll, demonstrated the combine harvester near Sarnia, Ontario, Canada. His self-propelled model was built in 1937 and demonstrated in 1938. The production of grain skyrocketed thereafter. The current epidemic of obesity, diabetes, heart disease, and autoimmune diseases was born.

Corn Combine Harvester

Top Twelve Historical Events

Event No. 11

Nathan Pritikin and Senator George McGovern

Nathan Pritikin developed a low-fat, low-protein, high-complex carbohydrate diet in the 1970s as a result of his obesity and clogged arteries. His program was based on a very low-fat, low-protein, low-calorie diet, and he was strictly against the use of any vitamins, minerals, or other supplements. The diet was able to reverse his obesity, but he committed suicide at the age of 60 after battling leukemia (cancer of the bone marrow). Studies have now shown that his low-fat, low-protein diet causes cancer. Others warned him that his diet had less vitamin E than minimum standards and that it was deficient in essential amino acids and essential fatty acids. His book was titled *Permanent Weight-Loss Manual*, and indeed he had a permanent weight loss—100%. Patients at his Longevity Center and the Pritikin Research Foundation who followed the diet without cheating are reported to have suffered multitudes of medical problems. Ann Louise Gittleman was his chief nutritionist.

Nathan Pritikin's diet program lowered total cholesterol, but new studies have shown that people with low cholesterol have an increased risk of suicide. Perhaps his low cholesterol combined with his leukemia contributed to his death by suicide.

Senator George McGovern attended the Pritikin clinic and was sold on the diet as the foundation for human nutrition. Upon returning to Washington, D.C., Senator McGovern used his Senate committee to investigate and recommend changes in nutrition. Because he was very biased, he blocked those scientists, physicians, and nutritionists who were opposed to the high-carbohydrate, low-fat, and low-cholesterol diet from testifying before the committee. This plan, which was passed on to other

Chapter 4

US government agencies without any hard scientific evidence to support it, became known as our present-day *USDA Food Guide Pyramid*. The last 30 years have proved this diet plan to be very hazardous to health, and it should never be followed by anyone. The United States government has spent hundreds of millions of dollars in research grants in their unsuccessful attempts to prove that saturated fat causes heart disease, only to find that those who eat the most saturated fat are the most active and healthy. Some fraudulent studies exist condemning saturated fat, but the real villain in the diet was ignored—carbohydrates.

The high-carbohydrate, low-fat *USDA Food Guide Pyramid* can be directly blamed on Nathan Pritikin and Senator George McGovern. The human pain, misery, disease, early death, and financial cost to societies around the world caused by Nathan Pritikin and Senator George McGovern's campaign against animal protein and animal fats are beyond measure. The climax to this dilemma is still years into the future. Millions of people in English-speaking countries—mainly the United States, Canada, United Kingdom, and Australia—have died early deaths from the complications of obesity, diabetes, heart disease, cancer, and numerous autoimmune diseases as a result of Senator McGovern's campaign. This global change in diet and health did not even require a line item description in the national budget. It became the official government diet recommendation with the hearty support of all the food manufacturers that turn carbohydrates and trans fats into **stuff** they call *food*.

Ann Louise Gittleman appeared to reverse her disastrous nutritional philosophy after Nathan Pritikin developed cancer and committed suicide. She described the Pritikin Diet failures in *Beyond Pritikin* and turned away from the low-fat diet in *Eat Fat, Lose Fat*.

Top Twelve Historical Events

Many people think the Atkins' low-carbohydrate diet is lacking essential nutrients because it doesn't line up with the *USDA Food Guide Pyramid*. Their reference is the US Food & Drug Administration's (USFDA) *Nutritional Guide for Daily Values (DV)* as shown on all nutrition labels. The US Department of Agriculture *Food Guide Pyramid* was developed by vegetarians with an agenda, and Nathan Pritikin and Senator George McGovern were the perpetrators. There is no science behind the *Food Guide Pyramid*. It was a scam from the beginning—a make-believe nutritional plan to limit the consumption of animal products. The results have been rampant heart disease, cancer, diabetes, osteoporosis, intestinal diseases, and a medical textbook full of other ailments. The *USFDA Nutritional Guide* is based on the *Food Guide Pyramid*. This is easy to verify. Simply go to a food count book and enter a 2000-calorie diet exactly according to the pyramid. The results will show every nutritional requirement to be perfectly met. It's all a scam. The DV was manipulated to fit the pyramid. There is no hard science behind the establishment of the USFDA daily nutritional requirements.

The health-care industries are about to enter a "meltdown" as health costs skyrocket from the onslaught of sickness brought on by the constant plea to "reduce the consumption of red meat and saturated fat." The 20 to 30-year delays in the onset of disease have begun to attack the "baby boomers." More and more people are unable to afford health-care insurance. Companies are dumping more and more of the health-care benefits back onto the employee. Health Maintenance Organizations (HMOs) are increasing rates at a pace three to five times the rate of inflation, but they still struggle to make a profit. Government health plans are heading for financial bankruptcy. Governments are frozen in their current policies, unable to reverse direction or fight the

Chapter 4

power of the food manufacturing lobby. The health-care crisis is still gathering momentum.

Meanwhile, pharmaceutical and medical equipment manufacturing companies are turning out new drugs and new machines that mainly serve to cover up the symptoms temporarily. These new products will only dump increased cost into the failing system. The healthcare crisis is destined to become more severe, not better. Obesity will continue increase as the government continues a "hard press" for people to adhere to the pyramid and get more exercise.

Soybean Irrigation System

Top Twelve Historical Events

Event No. 12

Soybean Revolution & GMO Foods

Just as grains became the major replacement for meat and fats in the diet, soy protein has become the vegetarian food of choice to replace meat protein. The natural animal fats had already been replaced by trans fats or vegetable oils. In place of fatty meats, vegetarians are now purchasing grain products bulked up with soy protein and held together by vegetable fats. Naturally, every package says "low-fat and low-cholesterol," which leads the unsuspecting shopper to believe these factory-made conglomerations are healthy foods.

Unfermented soy products contain enzyme inhibitors and unhealthy phytates that are not deactivated by cooking. Phytic acid combines with minerals in the intestines and prevents absorption. They are full of unhealthy isoflavones—the mega-doses of phytoestrogens that disrupt normal sex hormone and endocrine functions and lead to hypothyroidism. With all of these harmful ingredients, it is easy to understand why soy products contribute to many health problems.

Soybean plants growing beautifully in the fields are the most genetically modified organisms (GMO) in the United States' food supply. As of January 2006, the percentage of soybeans that has been genetically modified is estimated to be 75% and spreading fast. Virtually all products on supermarket shelves that contain any soy or soy protein have been contaminated. Just because a soy product is labeled *organic* does not mean it is GMO free. Vegetarians gorge themselves on soy products such as soy milk and soy burgers. Vegetarian foods, such as breads, bagels, cereals, crackers, pastas, and energy drinks, are often fortified with soy protein made from GMO-contaminated soybeans.

Chapter 4

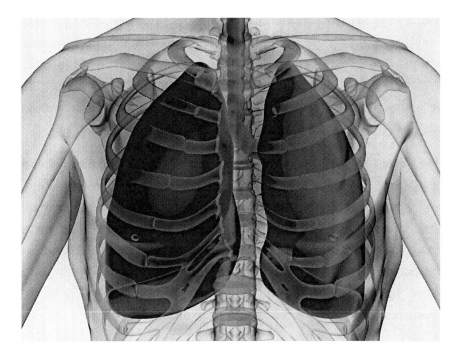

Asthma Is an Autoimmune Disease

**Remission Has Been Achieved by Following the
Perfect Diet—Perfect Nutrition
Presented in This Book**

Absolute Truth Exposed

Volume 1

Chapter 5

Achieving Remission in Asthma, Inflammatory Bowel Diseases, and Other Autoimmune Diseases with This New Diet Program

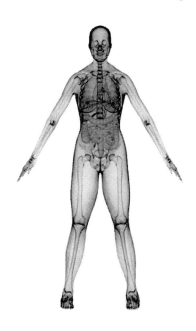

Chapter 5

This chapter will present detailed information about the following partial list of diseases classified as autoimmune: Asthma, Rheumatoid Arthritis (RA), Irritable Bowel Syndrome (IBS), Inflammatory Bowel Disease (IBD), Ulcerative Colitis (UC), Celiac, Crohn's Disease, Diverticulosis, Twisted and Redundant Colon, Proctitis, Fistulas, Fissures, Stomach and Duodenal Ulcers, Gastro-Esophageal Reflux Disease (GERD), Candida Albicans, Interstitial Cystitis, Sjogren's disease (dry eyes), Multiple Sclerosis (MS), Lupus Erythematosus, Fibromyalgia, Polymyalgia Rheumatica (PMR), Chronic Fatigue Syndrome (CFS), Psoriasis, and others.

Perfect Diet——Perfect Nutrition
Correct foods are almost miraculous.
Eating the wrong foods will cause a flare.

This chapter is not a medical approach to treating your autoimmune disease, although several prescription drugs are recommended for achieving the remission of symptoms more quickly than without them. This chapter is not strictly a nutritional approach to improving your autoimmune disease symptoms because prescription drugs may be required during flares or as recommended by your doctor. However, a reasonable remission is not possible without strict compliance with the nutrition guidelines presented. This chapter does not recommend naturopathic or Chinese medicine to heal your body, although several herbal products are recommended for specific purposes to achieve remission more quickly. As a scientist, I have taken an unrestricted scientific approach for achieving remission in autoimmune diseases, such as asthma, rheumatoid arthritis, ulcerative colitis, Crohn's disease, psoriasis, lupus, migraine headaches, multiple sclerosis, and several dozen other diseases. A cure is not possible as will be explained in more detail

Achieving Remission In Autoimmune Diseases

throughout the chapter. This book is all about science, not medicine.

You will be told to eat foods that other sources have told you to avoid, such as red meat, saturated animal fats, and hard cheeses; you will be told to strictly abstain from foods that other sources have told you to eat in abundance, such as gluten-free grains, high-fiber bran cereal, and fruit. Now is a good time to erase your brain and brace yourself for a direct attack on your nutritional dogma and addictions. Kiss the *USDA Food Guide Pyramid* goodbye forever. This fight requires your proactive charge against those forces that are keeping you sick. You will not be given a prescription for a drug and told to take two a day with meals. You should know that will not bring lasting relief. Get ready to win the battle in the same way that many others have. Let's get started.

I will present the **Starting Diet Overview** on page 235. The **Starting Diet** program (see page 245) is very restrictive in its elimination of foods that sufferers have identified as triggering adverse autoimmune responses. **Foods We Should Eat** (see page 266) is a less restrictive diet program for those sufferers who have achieved normal digestion and are without any major symptoms of the disease. This is called *remission*. You must make a lifetime commitment to refrain from the list of **Foods That Are Absolutely Forbidden**. (See page 270.) Eating only one bite of a forbidden food can cause an immediate autoimmune flare. Doctors and patients with autoimmune diseases such as asthma and rheumatoid arthritis are aware of the cycles between flares and remissions, but they don't know how to stop the flares. Remission is much different from a cure. My diet program stops the flares. **This is not a cure.** Your immune system has a very

Chapter 5

long memory and will attack your body again when you eat forbidden foods.

You may find the information here very interesting, but that does not necessarily translate into an improvement in your health. You must be totally proactive and make severe changes in diet and lifestyle in order to achieve remission of your autoimmune symptoms. It will never happen to those who have the attitude that they will go to the doctor and take the drugs he prescribes. Autoimmune diseases are not like hypothyroidism in which a little pill each morning is all it takes to resolve the problem.

I must issue a medical disclaimer at this point. This book will suggest many medical procedures and prescription medications. You must discuss them with your doctor before proceeding.

Medical Disclaimer

All information contained herein is intended for your general knowledge only and is not a substitute for medical advice or treatment for specific medical conditions. You should seek prompt medical care for any specific health issues and consult your physician before starting a new fitness or nutrition regimen. The information contained in this book is presented in summary form only and intended to provide broad consumer understanding and knowledge of diet, supplements, health, and diseases. The information should not be considered complete and should not be used in place of a visit, call, consultation, or advice from your physician or other health-care provider. I do not recommend the self-management of health problems. Do not self-diagnose your health condition based on the information provided in this book. Seek a diagnosis from your doctor. Information obtained by reading this book is not exhaustive and does not cover all diseases, ailments,

Achieving Remission In Autoimmune Diseases

physical conditions, or their treatment. Should you have any health-care related questions, please call or see your physician or other health-care provider promptly. You should never disregard medical advice or delay in seeking it because of something you have read here. I strongly suggest you select a physician who is knowledgeable and supportive of the low-carbohydrate lifestyle.

The diet program presented in this book is not to be interpreted as a cure for any disease. The diet information presented in this book is based on scientific analysis that show some foods cause negative reactions in some people and other foods do not.

I must emphasize that the criticism of some doctors, nutritionists, teachers, professors, scientists, literary agents, publishing companies, media companies, politicians, and others should not be viewed as a blanket criticism of all. Many truly awesome people in these industries are providing a valuable service.

Autoimmune diseases each exhibit very different symptoms which make them appear to be completely unrelated. However, they all have one commonality. They are all autoimmune disorders in which the immune system has been accidentally triggered to attack a specific protein tissue within the body. Your body is attacking itself. The original cause is commonly called *leaky gut syndrome*. Therefore, autoimmune diseases, such as asthma, rheumatoid arthritis, multiple sclerosis, lupus, Crohn's disease, ulcerative colitis, psoriasis, and all others are closely connected to inflammatory bowel diseases. For this reason, many people exhibit several of these diseases at the same time. If you suspect that you have an autoimmune disease, you should obtain a blood test for antinuclear antibodies.

Chapter 5

"Antinuclear Antibodies Test (ANA)"
http://www.medicinenet.com/antinuclear_antibody/article.htm

"What are antinuclear antibodies?"

"We normally have antibodies in our blood that repel invaders into our body, such as virus and bacteria microbes. Antinuclear antibodies (ANAs) are unusual antibodies, detectable in the blood, that have the capability of binding to certain structures within the nucleus of the cells. The nucleus is the innermost core within the body's cells and contains the DNA, the primary genetic material. ANAs are found in patients whose immune system may be predisposed to cause inflammation against their own body tissues. Antibodies that are directed against one's own tissues are referred to as auto-antibodies. The propensity for the immune system to work against its own body is referred to as autoimmunity. ANAs indicate the possible presence of autoimmunity and provide, therefore, an indication for doctors to consider the possibility of autoimmune illness."

"What other conditions cause ANAs to be produced?"

"ANAs can be produced in patients with infections (virus or bacteria), lung diseases (primary pulmonary fibrosis, pulmonary hypertension), gastrointestinal diseases (ulcerative colitis, Crohn's disease, primary biliary cirrhosis, alcoholic liver disease), hormonal diseases (Hashimoto's autoimmune Thyroiditis, Grave's disease), blood diseases (idiopathic thrombocytopenic purpura, hemolytic anemia), cancers (melanoma, breast, lung, kidney, ovarian and others), skin diseases (psoriasis,

Achieving Remission In Autoimmune Diseases

pemphigus), as well as in the elderly and those people with a family history of rheumatic diseases."

The high-carbohydrate, low-fat diet as outlined on the *USDA Food Guide Pyramid* causes leaky gut syndrome and dozens of autoimmune diseases. The low-fat, high-carbohydrate diet is inherently deficient in protein and essential fatty acids because people shun eating meat. This protein deficiency is widely ignored and denied by health officials, doctors, and professional nutritionists. The high-carbohydrate, low-fat, low-protein diet turns the intestines—especially the colon—into a fermentation tank consisting of a mass of pathogenic bacteria, yeasts, and parasites. Protein deficiency in the presence of these disease-causing pathogens degrades the lining of the intestines and leads to leaky gut syndrome. Food particles "leak" past the damaged lining of the intestines into the bloodstream where the immune system views them as foreign invaders. The immune system attacks these molecules, and because some of the leaky molecules are similar to healthy tissue, the body begins to attack itself. Typical autoimmune diseases caused by the high-carbohydrate diet are irritable bowel syndrome (IBS), inflammatory bowel disease (IBD), ulcerative colitis, Crohn's disease, diverticulosis, twisted and redundant colon, proctitis, fistulas, fissures, stomach and duodenal ulcers, interstitial cystitis, Sjogren's disease (dry eyes), rheumatoid arthritis, multiple sclerosis (MS), psoriasis, lupus erythematosus, asthma, and others.

Although appendicitis is not classified as autoimmune, it too is a result of the low-fat, high-carbohydrate diet. In addition, the deficiency of fat in the diet also prevents the gallbladder from discharging its contents, which results in gallstones and the

Chapter 5

possible surgical removal of the gallbladder known as a *cholecystectomy*.

Since doctors will typically prescribe drugs to treat each of the multiple autoimmune symptoms, people are led to believe that the diet should be different for each symptom. This is not the case. The diet shown to be beneficial for one autoimmune disease is the same for all of the diseases on the list. In their efforts to stop the autoimmune system attacks, doctors will also prescribe drugs to suppress the immune system. This is a bad idea on a long-term basis since most of these drugs have very harmful side effects. This diet program addresses the root cause by eliminating those foods that trigger the immune system to attack the body. Eating one bite from the **Forbidden Food** list can trigger an autoimmune flare that will last for weeks. Therefore, you must strictly comply with this diet program.

Warning --- Most people suffering from chronic intestinal problems are addicted to the very foods that made them sick and continue to prevent their recovery. These people typically refuse to change their nutritional philosophy. Please keep this in mind as you study the diet suggestions presented here and compare them to the foods you eat. Breaking out of your addictions and nutritional beliefs can be extremely difficult.

Add brainwashing to the list of obstacles or barriers that you must overcome in order to be successful in using this diet program to relieve the autoimmune disease symptoms.

A great many people don't even get to first base. They strike out without ever attempting to try the diet program. Perhaps the following testimonies will help you overcome all of these difficult obstacles. Others have succeeded in ways that are almost miraculous. You can succeed as well. Don't give up.

Achieving Remission In Autoimmune Diseases

Recent Testimonies from People with Ulcerative Colitis, Crohn's Disease, and Candida

Sorin's doctor had given him many drugs that did not make him well. The doctor gave up and sent him home to suffer. After he had been on my diet program for several months, Sorin wrote from Romania on November 23, 2009:

"The diet plan plus supplements saved my life."

"I thought I would sign in for a progress report! I am doing very well. So far, so good. I am down to 2 BM's a day, and bleeding has disappeared. I am eating a very balanced diet from your Starting Diet and it seems as if I am able to tolerate a lot of the red meat now. Thank you for all your advice and I hope all is well with you."

Note: Sorin wrote again on August 12, 2008:

"I told my doctor that your diet had made me well. He was very angry and said, 'Get out of my laboratory.' Anyway, I can tell you that I cannot have a normal life anymore here. Everyone wants to put me in a psychiatric hospital because of your diet. I'm healing and growing, being healthy, and they want to put me in a hospital. I usually eat 4 times a day mainly meat, fish, fowl, and eggs. I'm perfectly well and they want to put me in a nerve hospital. Kent, Please ask your God to stop everyone to make me trouble. I was on your diet and everything was fine. So what if I eat hard cheese, eggs, beef, lamb, and pork."

Note: The following report tells of a Crohn's colitis healing that occurred in only a few months on my diet program.

Chapter 5

"I had colitis problems since 2 years of age. Chronic constipation was a problem for 30 years as an adult. I was diagnosed as having Crohn's colitis disease three years ago when the constipation changed to diarrhea. After a few months on your diet I am now free of all symptoms, and I discontinued my drug. My recent colonoscopy showed a perfectly normal colon. My doctor was amazed. The arthritis in my hands had been causing stiffness and pain. It was getting worse each year but has completely disappeared. I can knit again."

Note: The first report by Sorin and the report below from Sweden are typical of doctors' denial. Don't expect your doctor to agree with this diet program. Most likely your doctor will strongly object, and he certainly won't give the diet any credit for your healing.

"I have been following your diet for a couple of months. I eat mostly moose meat with coconut fat. I had a colonoscopy three days ago and guess what? No inflammation! My colon looked so good my doctor said the previous inflammation could not have been from a flare. Yeah, right! I know it was a bad flare, but I am fine now. Your page was the first site to open my eyes to the dangers of eating carbs. Keep up the good work."

Note: The following is a three-month update.

"Just wanted to give you another update. My asthma is all gone. I'm eating 4 lbs of red meat every day, and I'm growing like a weed! Shredding fat, adding muscle and being healthy is the best thing that has ever happened to me!"

Achieving Remission In Autoimmune Diseases

Note: The report below tells how my diet program works when other diets have failed.

"Thanks for your information, it's been very helpful. I've taken your advice, and I've eliminated many of the offending foods that were allowed on another type of IBD diet that I thought were alright. That diet includes yogurt, melon, goat's cheese and surprisingly, the most helpful of all carrots! I've gone almost a year thinking that I couldn't tolerate butter and fatty meat, and suddenly they cause me no problem. I was having a flare for three weeks, and just beginning with the principals of your diet (not including the supplements and minerals) has gotten me back to normality in less than a week. It's a wonder!"

Note: The ex-vegetarian below experienced awesome health improvements when he switched to my high-fat, red meat diet. The vegetarian diet caused his candida infection, leaky gut syndrome, and intestinal and degenerative disc diseases.

"Dear Mr. Rieske,

How can I ever thank you? Your knowledge, insight, honesty, and courage contained in your articles have significantly improved the quality of my life!

I am now 37 years old. From the ages of 25 to 36 I had adopted a predominantly vegetarian lifestyle. Over the years I have developed a mild case of degenerative disc disease, constant headaches, and a severe case of Candida (and perhaps leaky gut). I started to figure out that something was definitely wrong. The mainstream nutritional advice had definitely steered me in the wrong direction.

Chapter 5

I have been eating a "clean" and "natural" diet for several months now on red meat, fish, fowl, a small amount of veggies and berries (I can't help it, I love berries). My headaches have disappeared! I went from 172 lbs (at 5'8" tall) to 158 lbs! I have no constipation nor diarrhea (which was at one time constant)! I think clearly, my mood is better, and I feel great!

You have made a huge impact on my life, and I am grateful. I have discussed your information with every person in my social circle as it is all I can seem to do to help others as you've helped me.

Thanks again, and God Bless,

Stuart"

Note: Fabien writes from France on July 9, 2010.

"Dear Kent,

I'm still doing the starting diet, now from March. Many things are going better so I'm fully convinced that carbs are poison. Thanks for your incredible work and analysis on the right nutrients made for human.

Just a remark: last week, as I felt good (eg. full energy for my job), and I've eaten some forbidden foods like chocolate, cake at my cousin's birthday, and a lunch into a restaurant for my job. I thought I would be ok with these exceptions as I felt better, but unfortunately the week end and this week I have felt again bad symptoms like spondylarthris crisis, excessive tiring, heavier pain on colon... So it seems that even once you think you are cured, you cannot eat some forbidden foods yet.

Thanks for reading me, I feel so alone in France! All people are against me regarding this diet (doctor, parents, friends....). They don't want to understand that I feel much better now.

Best wishes dear Kent, God bless you.

Fabien"

These testimonies show that in a very short time my diet program as presented in this book will restore good health to those suffering from a variety of autoimmune diseases. I receive a common report from people who attempt to tell their doctor about this diet and the awesome benefits. Many doctors get angry and threaten to dismiss these patients from their care. Sorin's doctor told him, "Get out of my laboratory." Physicians in English-speaking countries would have said "office" instead of "laboratory" as is the normal reference in Sorin's country. Don't expect your doctor to support or agree with my diet program. Most likely you will receive criticisms, threats, and warnings that you are putting your health at a greater risk.

Chapter 5

Descriptions of Common Autoimmune Diseases and the Official Medical Treatments

The following descriptions of several common autoimmune diseases are generally accepted. The causes for these diseases are not identified because professional medical associations do not accept leaky gut syndrome as an explanation for these conditions.

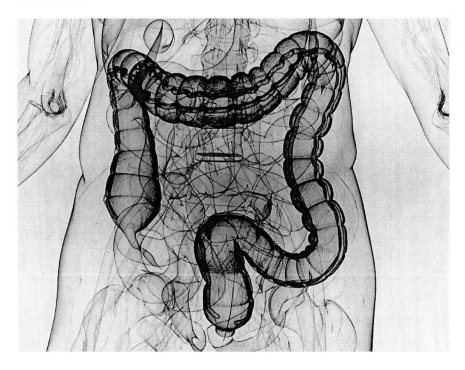

A Healthy Colon Is the Key to Avoiding Autoimmune Diseases

The official medical treatment for autoimmune disease is often trial and error, and since the results are limited, it has become standard practice to suppress the immune system with drugs. Although improvement can be seen during the first few weeks, it declines over time. These immunosuppressant drugs have the potential for severe side effects, the most common of which is

Achieving Remission In Autoimmune Diseases

degenerative disc disease. It is so debilitating that it can be as bad as or worse than the autoimmune disease itself. President John Kennedy suffered this fate. Immunosuppressant drugs also increase the risk of deadly infections.

Diet recommendations given by the doctor are generally ineffective because they are either not strict enough or they restrict the wrong foods. Doctors sometimes tell patients with asthma to avoid grains that contain gluten. Although this may provide some improvement in the severity of the symptoms, the symptoms still remain. The gluten-free diet may not help at all because many offensive foods are still being consumed.

"What is Irritable Bowel Syndrome?"
http://digestive.niddk.nih.gov/ddiseases/pubs/ibs/index.htm

> "Irritable bowel syndrome is a disorder characterized most commonly by cramping, abdominal pain, bloating, constipation, and diarrhea. IBS causes a great deal of discomfort and distress, but it does not permanently harm the intestines and does not lead to a serious disease, such as cancer. Most people can control their symptoms with diet, stress management, and prescribed medications. For some people, however, IBS can be disabling. They may be unable to work, attend social events, or even travel short distances."
>
> "As many as 20% of the adult population, or one in five Americans, have symptoms of IBS, making it one of the most common disorders diagnosed by doctors. It occurs more often in women than in men, and it begins before the age of 35 in about 50% of people."
>
> **What is the treatment for IBS?**

Chapter 5

"Unfortunately, many people suffer from IBS for a long time before seeking medical treatment. Up to 70% of people suffering from IBS are not receiving medical care for their symptoms. No cure has been found for IBS, but many options are available to treat the symptoms. Your doctor will give you the best treatments for your particular symptoms and encourage you to manage stress and make changes to your diet."

"Nutrition and Inflammatory Bowel Disease" *by InteliHealth.*
http://www.intelihealth.com/

"Inflammatory bowel disease is the term used for both Crohn's disease and ulcerative colitis. These conditions may be associated with malnutrition because of such factors as poor appetite, decreased absorption, surgical loss of bowel, intestinal injury and chronic diarrhea. Patients commonly experience weight loss. Vitamin and trace-element deficiency may also occur. If the disorder occurs during childhood, growth failure due to malnutrition must be treated before the potential for normal growth is lost."

"Diet can't cure inflammatory bowel disease. Avoiding certain foods, however, may help to control symptoms. Foods to avoid include dairy foods high in lactose, such as milk and soft cheeses (in patients who are lactose intolerant), raw fruits and vegetables, spicy foods and nuts."

"Specialized, predigested liquid diets (elemental diets), combined with the exclusion of other food, may induce remission in some patients. Recent data suggest less

Achieving Remission In Autoimmune Diseases

highly predigested (polymeric) diets may work as well as elemental diets. Preliminary studies have examined the benefits of using fish oils to relieve the symptoms of ulcerative colitis or prolong remission in Crohn's disease."

Irritable Bowel Syndrome
http://digestive.niddk.nih.gov/ddiseases/pubs/ibs/index.htm

"Irritable bowel syndrome (IBS) is a common but poorly understood disorder that causes a variety of bowel symptoms including abdominal pain, diarrhea and/or constipation, bloating, gassiness and cramping. While these symptoms may be caused by a number of different bowel diseases, IBS is usually diagnosed only after your doctor has ruled out the possibility of a more serious problem. The severity of the disorder varies from person to person. Some patients experience intermittent symptoms that are just mildly annoying, while others may have such severe daily bowel problems that IBS affects their ability to work, sleep and enjoy life. In addition, symptoms may change over time, such that an individual may have severe symptoms for a several weeks and then feel well for months, or even years. Most people are never cured of IBS, but the disorder is not related to and does not progress to any other disease, such as ulcerative colitis or colon cancer."

Chapter 5

Celiac Disease
http://digestive.niddk.nih.gov/ddiseases/pubs/celiac/index.htm

"Celiac disease (also called non-tropical sprue, celiac sprue and gluten-sensitive enteropathy) is an intestinal disorder in which gluten—a natural protein commonly found in many grains, including wheat, barley, rye and oats—cannot be tolerated by the body. In normal, healthy people, gluten is digested like any other protein or nutrient, then absorbed through the fingerlike protrusions called villi (and the even tinier hairlike protrusions called microvilli) that make up the surface of the small intestine. But in people with celiac disease, gluten causes an immune reaction in the body, leading to inflammation and damage to the villi and microvilli of the small intestine. The result is that the intestine cannot absorb nutrients properly and the person can become malnourished."

"Celiac disease is believed to be hereditary and is most common among people of northern European descent. As many as one in every 250 Americans may have celiac disease, but most cases go undiagnosed, often because the symptoms are assumed to be caused by other diseases or because symptoms are mild. Although many cases are diagnosed in childhood, celiac disease can be diagnosed at any age. It is believed that surgery, pregnancy, childbirth, viral infection or severe emotional stress can trigger symptoms of the disease in people who are predisposed genetically."

"Celiac disease is considered to be an autoimmune condition because the body's own immune system damages the intestinal villi, even though the process is

Achieving Remission In Autoimmune Diseases

started by eating gluten. People with celiac disease also tend to have an increased risk of developing other autoimmune diseases, such as type 1 diabetes (which usually requires insulin and is most common in children and young adults), systemic lupus erythematosus, rheumatoid arthritis and thyroid disease. Other conditions that may coexist with celiac disease include dermatitis herpetiformis (an itchy, blistering skin rash), liver disease and immune system abnormalities."

February 11, 2003 05:39 P.M. ET By Alison McCook
NEW YORK (*Reuters Health*)

"New research shows that some people with no obvious symptoms of the digestive disorder celiac disease can, in fact, have the condition. A survey of more than 4,000 people with no genetic risk factors or symptoms of celiac disease revealed that about 1 in 130 had the disorder. Over time, celiac disease can damage the small intestine and interfere with the normal absorption of nutrients from food. The only treatment for the illness is a diet that strictly avoids foods that contain gluten, a protein found in wheat, rye and barley."

Ulcerative Colitis
http://digestive.niddk.nih.gov/ddiseases/pubs/colitis/index.htm

"Ulcerative colitis is a lifelong condition that begins with rectal inflammation but can progress to involve much or all of the large intestine. Ulcerative colitis typically begins to cause symptoms in young adulthood, most often between the ages of 15 and 40."

Chapter 5

"No one knows for sure what triggers the inflammation in ulcerative colitis. Many people think that it begins with a virus or a bacterial infection and that the body's immune system malfunctions and stays active after the infection has cleared. In this kind of autoimmune problem, the bowel is injured by the body's immune system. The disease does not act contagious, even within families, so there is no worry of direct spread of the disease from one person to another."

"Ulcerative colitis affects the inner lining of the rectum and adjoining colon, causing it to wear away in spots (leaving ulcers), to bleed or to ooze cloudy mucus or pus. A few other parts of the body seem to be affected by inflammation and can develop symptoms in ulcerative colitis, including the eyes, skin, liver, back and joints."

Crohn's Disease
http://digestive.niddk.nih.gov/ddiseases/pubs/crohns/index.htm

"Crohn's disease is a chronic condition associated with inflammation and injury of the intestines. It typically begins to cause symptoms in young adulthood, most often between 15 and 40 years of age."

"No one knows for sure what triggers the intestinal inflammation. Many people think that a virus or a bacterial infection might start the process, and that the body's immune system "malfunctions" and stays active after the infection has cleared. Family members may share genes that make Crohn's disease or another similar inflammatory bowel disease, ulcerative colitis, more likely to develop if the right trigger occurs. (20% to 25% of people who have Crohn's disease have at least one relative with Crohn's or ulcerative colitis.) The disease is not contagious, even within families, so there is no worry of direct spread of the disease from one person to another."

"Once Crohn's disease begins, it can cause intermittent, lifelong symptoms by inflaming the inside lining and deeper layers of the intestine wall. The irritated intestine lining can thicken or wear away in spots (creating ulcers) or in cracks (creating fissures). Inflammation can also allow an abscess (a pocket of pus) to develop. In between attacks of inflammation, the intestine attempts to heal by recoating itself with a new inside lining. When the inflammation has been severe, the intestine can lose its ability to distinguish the inside of one piece of intestine from the outside of another piece. As a result, it can mistakenly build a lining along the edges of an ulcer that has worn through the whole wall of the gut, creating a fistula—a permanent tunnel between one piece of gut and another. A fistula can sometimes even form between the gut and the skin surface, creating drainage of mucous to the skin."

Diverticulosis
http://digestive.niddk.nih.gov/ddiseases/pubs/diverticulosis/index.htm

"Diverticulosis is a colon condition so common among older adults in developed countries -- and so rare in rural Africa and Asia -- that it has been called " a disease of Western civilization." Diverticulosis develops in weak spots along the walls of the colon, typically in the sigmoid colon on the left side of the abdomen. In response to pressure, small, balloonlike pouches called diverticula

Chapter 5

may poke outward through the weak spots. Most people have no symptoms, but if a diverticulum bleeds or becomes infected, hospital care generally is required. Surgery may be recommended in specific circumstances."

The diverticula can become infected and inflamed by pathogenic bacteria. A combination of two prescription drugs is commonly used for diverticulitis—Flagyl® (metronidazole) plus Levofloxacin. Unfortunately, the drugs kill off the good intestinal floras that are needed to block the return of the opportunistic pathogens. The best approach is to start the diet program presented here and begin the drugs at the same time. Increase the probiotics to two capsules each with every meal for one month. Take probiotics at the same time as the drugs and continue to take probiotics after the drug program has ended.

Some publications suggest that Mycobacterium paratuberculosis causes Crohn's disease. Many opportunistic pathogenic bacteria appear in Crohn's patients because their digestions are not normal and they are taking immunosuppressant drugs. These drugs bring temporary relief from the symptoms, but suppressing the immune system allows the Mycobacterium to multiply. Killing these bacteria provides temporary relief but not a permanent eradication. The high-carbohydrate, high-fiber, low-fat diet recommended by most gastroenterologists provides the opportunity for the Mycobacterium paratuberculosis to return and flourish. This is why Crohn's patients' conditions continue to deteriorate over the years. The diet program presented here stops this escalating cycle by removing the foods that feed the bacteria—carbohydrates and fiber. It reduces the negative immune system response to carbohydrates and fiber and increases the strength of the immune system to fight the

Achieving Remission In Autoimmune Diseases

Mycobacterium. The "Perfect Diet" program presented here brings a better result.

"Handout on Health: Rheumatoid Arthritis"
http://www.niams.nih.gov/Health_Info/Rheumatic_Disease/default.asp

What is Rheumatoid Arthritis?

"Rheumatoid arthritis is an inflammatory disease that causes pain, swelling, stiffness, and loss of function in the joints. It has several special features that make it different from other kinds of arthritis. For example, rheumatoid arthritis generally occurs in a symmetrical pattern, meaning that if one knee or hand is involved, the other one also is. The disease often affects the wrist joints and the finger joints closest to the hand. It can also affect other parts of the body besides the joints. In addition, people with rheumatoid arthritis may have fatigue, occasional fevers, and a general sense of not feeling well."

"Rheumatoid arthritis affects people differently. For some people, it lasts only a few months or a year or two and goes away without causing any noticeable damage. Other people have mild or moderate forms of the disease, with periods of worsening symptoms, called flares, and periods in which they feel better, called remissions. Still others have a severe form of the disease that is active most of the time, lasts for many years or a lifetime, and leads to serious joint damage and disability."

How Is Rheumatoid Arthritis Treated?

"Although rheumatoid arthritis can have serious effects on a person's life and well-being, current treatment

Chapter 5

strategies--including pain-relieving drugs and medications that slow joint damage, a balance between rest and exercise, and patient education and support programs--allow most people with the disease to lead active and productive lives. In recent years, research has led to a new understanding of rheumatoid arthritis and has increased the likelihood that, in time, researchers will find even better ways to treat the disease."

"NINDS Multiple Sclerosis Information Page"
http://www.ninds.nih.gov/disorders/multiple_sclerosis/multiple_sclerosis.htm

What is Multiple Sclerosis?

"An unpredictable disease of the central nervous system, multiple sclerosis (MS) can range from relatively benign to somewhat disabling to devastating, as communication between the brain and other parts of the body is disrupted. Many investigators believe MS to be an autoimmune disease -- one in which the body, through its immune system, launches a defensive attack against its own tissues. In the case of MS, it is the nerve-insulating myelin that comes under assault. Such assaults may be linked to an unknown environmental trigger, perhaps a virus."

"Most people experience their first symptoms of MS between the ages of 20 and 40; the initial symptom of MS is often blurred or double vision, red-green color distortion, or even blindness in one eye. Most MS patients experience muscle weakness in their extremities and difficulty with coordination and balance. These symptoms may be severe enough to impair walking or even standing.

Achieving Remission In Autoimmune Diseases

In the worst cases, MS can produce partial or complete paralysis. Most people with MS also exhibit paresthesias, transitory abnormal sensory feelings such as numbness, prickling, or "pins and needles" sensations. Some may also experience pain. Speech impediments, tremors, and dizziness are other frequent complaints. Occasionally, people with MS have hearing loss. Approximately half of all people with MS experience cognitive impairments such as difficulties with concentration, attention, memory, and poor judgment, but such symptoms are usually mild and are frequently overlooked. Depression is another common feature of MS."

Is There Any Treatment?

"There is as yet no cure for MS. Many patients do well with no therapy at all, especially since many medications have serious side effects and some carry significant risks. In the past, the principal medications used to treat MS were steroids including adrenocorticotropic hormone (better known as ACTH), prednisone, prednisolone, methylprednisolone, betamethasone, and dexamethasone. While steroids do not affect the course of MS over time, they can reduce the duration and severity of attacks in some patients. Spasticity, which can occur either as a sustained stiffness caused by increased muscle tone or as spasms that come and go, is usually treated with muscle relaxants and tranquilizers such as baclofen, tizanidine, diazepam, clonazepam, and dantrolene. Physical therapy and exercise can help preserve remaining function, and patients may find that various aids -- such as foot braces, canes, and walkers -- can help them remain independent and mobile. If psychological symptoms of fatigue such as

Chapter 5

>depression or apathy are evident, antidepressant medications may help."

A deficiency of cholesterol in the body and the diet has been found to cause malformation of the myelin sheath of the nerves, central nervous system, and brain. This could be a contributing factor in multiple sclerosis—autoimmune disease of the myelin sheath.

As of January 2010, the United States has 400,000 cases of multiple sclerosis, or 16% of the 2,500,000 cases worldwide; yet the population of the United States is only 4.6% of the total world population. The *USDA Food Guide Pyramid* and massive government efforts to promote the pyramid diet have increased the incidence of many diseases, including multiple sclerosis.

"Handout on Health: Systemic Lupus Erythematosus"
http://www.niams.nih.gov/Health_Info/Lupus/default.asp

What is Lupus?

>"Lupus is one of many disorders of the immune system known as autoimmune diseases. In autoimmune diseases, the immune system turns against parts of the body it is designed to protect. This leads to inflammation and damage to various body tissues. Lupus can affect many parts of the body, including the joints, skin, kidneys, heart, lungs, blood vessels, and brain. Although people with the disease may have many different symptoms, some of the most common ones include extreme fatigue, painful or swollen joints (arthritis), unexplained fever, skin rashes, and kidney problems."

Achieving Remission In Autoimmune Diseases

What is the Treatment for Lupus?

"At present, there is no cure for lupus. However, lupus can be effectively treated with drugs, and most people with the disease can lead active, healthy lives. Lupus is characterized by periods of illness, called flares, and periods of wellness, or remission. Understanding how to prevent flares and how to treat them when they do occur helps people with lupus maintain better health. Intense research is underway, and scientists funded by the NIH are continuing to make great strides in understanding the disease, which may ultimately lead to a cure."

"What Causes Asthma?"
National Heart, Lung, and Blood Institute.
http://www.nhlbi.nih.gov/health/dci/Diseases/Asthma/Asthma_Causes.html

"The exact cause of asthma isn't known. Researchers think a combination of factors (family genes and certain environmental exposures) interact to cause asthma to develop, most often early in life. These factors include:

- An inherited tendency to develop allergies, called atopy (AT-o-pe)
- Parents who have asthma
- Certain respiratory infections during childhood
- Contact with some airborne allergens or exposure to some viral infections in infancy or in early childhood when the immune system is developing

If asthma or atopy runs in your family, exposure to airborne allergens (for example, house dust mites, cockroaches, and possibly cat or dog dander) and irritants

Chapter 5

(for example, tobacco smoke) may make your airways more reactive to substances in the air you breathe.

Different factors may be more likely to cause asthma in some people than in others. Researchers continue to explore what causes asthma."

How Is Asthma Treated and Controlled?

"Asthma is a long-term disease that can't be cured. The goal of asthma treatment is to control the disease. Good asthma control will:

- Prevent chronic and troublesome symptoms such as coughing and shortness of breath
- Reduce your need of quick-relief medicines (see below)
- Help you maintain good lung function
- Let you maintain your normal activity levels and sleep through the night
- Prevent asthma attacks that could result in your going to the emergency room or being admitted to the hospital for treatment

To reach this goal, you should actively partner with your doctor to manage your asthma or your child's asthma. Children aged 10 or older—and younger children who are able—also should take an active role in their asthma care.

Taking an active role to control your asthma involves working with your doctor and other clinicians on your health-care team to create and follow an asthma action plan. It also means avoiding factors that can make your asthma flare up and treating other conditions that can interfere with asthma management."

Achieving Remission In Autoimmune Diseases

Sjogren's Dry Eye Syndrome
http://www.intelihealth.com/

"Sjogren's syndrome is a chronic autoimmune disorder in which the body's immune defenses attack the salivary glands, the lacrimal glands (glands that produce tears), and occasionally the skin's sweat and oil glands. In some cases, the illness also affects the lungs, liver, vagina, pancreas, kidneys and brain. Most people with this disease are women who first develop symptoms during middle age. In about 50% of cases, the illness occurs together with rheumatoid arthritis, systemic lupus erythematosus (lupus), scleroderma or polymyositis."

What Causes Psoriasis
http://www.intelihealth.com

"Psoriasis is a chronic (long-lasting) skin disorder that causes scaling and inflammation. Psoriasis affects 2% to 3% of all people. It may develop as a result of an abnormality in the body's immune system, which normally fights infection and allergic reactions. Psoriasis probably involves heredity, because up to 40% of patients have family members with the same problem. Certain medications, such as lithium, a medication for bipolar disorder, may trigger psoriasis. Other medications, including beta-blockers, a class of heart and blood pressure medicines, seem to make psoriasis worse in people who already have the disease."

What are the Symptoms?

"Psoriasis causes skin scaling and inflammation, with or without itching. There are several types of psoriasis:

Chapter 5

- In **plaque psoriasis**, there are rounded or oval patches (plaques) of affected skin. These are usually red and covered with a thick silvery scale. The plaques often occur on the elbows, knees, scalp or near the buttocks. They may also appear on the trunk, arms and legs.
- **Inverse psoriasis** is a plaque type of psoriasis that tends to affect skin creases, especially those in the underarm, groin, buttocks, genital areas or under the breast. The red patches of inverse psoriasis may be moist rather than scaling.
- In **pustular psoriasis**, the skin patches are studded with pimples or pustules.
- In **guttate (meaning droplike) psoriasis**, many dime-sized or smaller red, scaly patches develop suddenly and simultaneously, often in a young person who has recently had a strep throat or a viral upper respiratory infection.

About 50% of people with skin symptoms of psoriasis also have abnormal fingernails, especially nail thickening or small indentations, called pitting. A type of arthritis, called psoriatic arthritis, can affect 10% to 20% of all people with psoriasis, and in some people, it occurs before skin changes appear."

How is Psoriasis Treated?

"Treatment for psoriasis varies depending on the type of psoriasis, the amount and location of affected skin, and the risks and benefits of each type of treatment.

- **Topical treatments.** These are treatments applied directly to the skin. They include daily skin care with emollients for lubrication, such as petroleum jelly or unscented moisturizers. Corticosteroid creams, lotions and ointments may be prescribed in medium

and high-potency forms for applying to stubborn plaques on the hands, feet, arms, legs and trunk, and may be prescribed in low-potency forms for areas of delicate skin such as the face.
- **Phototherapy.** For extensive or widespread psoriasis, light treatment, using ultraviolet B or ultraviolet A, may be used alone or combined with coal tar. A treatment called PUVA (psoralin + UVA) combines ultraviolet A light treatment with an oral medication called psoralen, which improves the effectiveness of the light treatment. Laser treatment also can be used. It allows treatment of the involved skin to be more focused so that higher amounts of UV light can be used.
- **Treatments for moderate to severe psoriasis involving large areas of the body.** These include vitamin A derivatives -- acitretin (Soriatane), methotrexate (Folex, Methotrexate LPF, Rheumatrex) -- and cyclosporine (Neoral, Sandimmune). These treatments are very powerful, and some have the potential to cause severe side effects involving the liver, kidney or blood. Therefore, it's essential to understand the risks and be monitored closely.

Psoriasis is an autoimmune disease in which the body's immune system is attacking the skin and other tissues. Standard medical treatments attempt to minimize the symptoms. My diet program uses a scientific nutritional approach by providing all of the essential nutrients in abundance and eliminating those foods that cause the immune system to attack the body. However, standard medical treatment can be used at the same time.

Chapter 5

Lack of Omega-6 Arachidonic Fatty Acid Associated with Severe Dermatitis

University of Illinois – April 14, 2010

"A specific omega-6 fatty acid may be critical to maintaining skin health, according to University of Illinois scientists."

"In experiments with mice, we knocked out a gene responsible for an enzyme that helps the body to make arachidonic acid. Without arachidonic acid, the mice developed severe ulcerative dermatitis. The animals were very itchy, they scratched themselves continuously, and they developed a lot of bleeding sores," said Manabu Nakamura, a U of I associate professor of food science and human nutrition."

"When arachidonic acid was added to the animals' diet, the itching went away, he said."

Does your skin itch? Eat some high-fat meat. Pork shoulder roast is high in arachidonic acid.

Achieving Remission In Autoimmune Diseases

What Do Doctors Say about Autoimmune Diseases?

The average doctor will tell you that Crohn's disease, arthritis, multiple sclerosis, psoriasis, lupus, asthma, and other autoimmune diseases have no known cause and cannot be healed. Many gastroenterologists will say that your diet did not cause your inflammatory bowel disease and that a change in your diet will not help your condition. They may recommend a high-fiber diet, which actually makes inflammatory bowel diseases worse. High-fiber foods are also high-carbohydrate foods. Humans do not have the enzyme needed to digest fiber, but the bad pathogenic bacteria in the colon certainly do. Fiber will make your IBD worse—guaranteed. On the other hand, good intestinal bacteria thrive on dietary saturated and monounsaturated fats like those found in red meat. The bad, disease-causing intestinal bacteria thrive on carbohydrates and fibers found in fruit, whole grains, raw vegetables, and legumes.

I received the following comment from the mother of a young son who has been suffering from IBD. His gastroenterologist (her nickname for the doctor is *Gastro*) prescribed two common immunosuppressant drugs for his treatment. Although the boy still remained very ill while taking the drugs, his digestive health began to improve immediately upon starting my diet program, and his mother began to taper off the drugs with the concurrence of his doctor. Because the mother was new to the diet, she was not sure why her son was improving until she followed the doctor's recommendation that the boy add custard, ice cream, and bread to his diet. The doctor had recommended "no change to his diet" but proceeded to change it. Adding these **Forbidden Foods** caused an immediate flare as she vividly describes. Her experience clearly shows that gastroenterologists are not properly trained to provide nutritional treatment for patients

Chapter 5

with IBD. I also recommended she keep a journal (she calls it a diary) to record health symptoms and diet information to identify food reactions. This is her testimony.

> "Just a note to you - The Gastro recommended no changes in his diet adding that custard, ice cream plenty of bread etc was okay - BAD MOVE - He was sick again and the symptoms returned when the Prednisone was stopped - so that logically tells me that your diet is paramount. Also now we have a diary. I am very aware that milk and sugars of any kind give him stomach ache. Thanks for your help."

My diet program allows a few specific vegetables but they must be thoroughly cooked. Cooking breaks down the fiber into simple carbohydrates that the body digests in the small intestines, thereby preventing the fiber from reaching the colon where it can become food for pathogens. The list of acceptable vegetables was derived by observing people who suffer with IBD. The exact reason why some vegetables are satisfactory and some exceedingly harmful is unknown.

The major medical professions and the American Medical Association (AMA) generally claim that carbohydrates are innocent and necessary dietary requirements. In fact, the scientific and nutritional requirement for carbohydrates in the diet is zero—none. Most professional nutritionists and the *USDA Food Guide Pyramid* falsely claim that carbohydrates are required in the diet. Studies have consistently shown that the lower the dietary intake of carbohydrates, the better the long-term health.

The entire family of related inflammatory bowel diseases is most likely caused by the excessive consumption of carbohydrates in combination with a deficiency of protein

Achieving Remission In Autoimmune Diseases

and fat from animal sources. Many people with IBD have a common dietary history of shunning meat and natural animal fats, particularly red meat and saturated fat.

The very high rate of intestinal diseases among vegetarians is the "dirty little secret" of nutritional health. Vegetarians in Southern China and Southern India have a very high incidence of IBD and colon cancer. Internet chat rooms, message boards, and forums for intestinal disease topics are packed with vegetarians who are hopelessly locked into their dietary religion. They suffer from a continuous worsening of inflammatory bowel diseases because of these sacred dietary restrictions. The percentage of vegetarians in the general population of the US is about 5%, but the percentage of vegetarians and ex-vegetarians on IBD message boards is about 80%. One doesn't need a double-blind study to see the obvious, but one certainly must be doubly blind to deny it. Colon cancer among vegetarians is epidemic, but the media keep a tight lid on the topic. People are continually falling away from the vegetarian diet because they realize it is harming their health.

> **There is an epidemic of young college women who have already blown their guts out eating the vegetarian diet.**

The vegetarian diet among college women is growing and producing disastrous results. They suffer persistent constipation or diarrhea because of a diet of fruit, whole grain cereals, bagels, beans, rice, tofu, and soy milk. They stop menstruating and have hair loss and dry skin problems. Jogging, drinking high-energy veggie drinks, and eating energy bars make it worse.

More women than men tend to shun red meat and saturated fat as advised by the *USDA Food Guide Pyramid*. Women may eat some chicken and fish but in insufficient quantities. More women

Chapter 5

than men have been brainwashed into believing red meat and saturated fats are unhealthy. Some women have become so brainwashed they say the thought of eating red meat and saturated fat "makes them gag." They substitute carbohydrates for red meat and saturated fat which results in a much higher incidence of autoimmune diseases among women than among men.

Candida Albicans Yeast and Fungus Systemic Overgrowth Infections

A candida albicans yeast, filamentous fungi, or dimorphic fungi infection in the small intestine or colon occurs frequently in people with inflammatory bowel diseases. The excessive consumption of carbohydrates and sugars causes candidiasis (a systemic candida albicans yeast fungus overgrowth). The yeasts produce alcohol as they consume the carbohydrates and sugars. The infected person becomes addicted to the alcohol and experiences withdrawal systems when the sweets are restricted. In order to eliminate the yeast infection, carbohydrates and sugars must be restricted. Because the dying yeasts give off toxins, the low-carbohydrate diet will initially make the infected person feel worse. Although eating sweets and sugars may make the person feel better, they simply feed the yeast and promote the spread of the infection.

Fungi or toxic molds in foods such as grains produce mycotoxins and metabolites which cause allergies and diseases. Gliotoxins (fungal toxins produced in the body by various species of Trichoderma, Gladiocladium fimbriatum, Aspergillus fumigatus, and Penicillium) have a severe immunosuppressive effect. Candida places a heavy load on the immune system which leads to the onset of other diseases. People who become vegetarians or those who avoid eating meat and fats are most likely to develop a

Achieving Remission In Autoimmune Diseases

systemic candida albicans yeast or fungal infection because of the over-consumption of carbohydrates and fiber. The symptoms can be subtle, and the diagnosis can be easily missed without a laboratory stool test.

The diet program presented here suppresses candida albicans yeast and fungus in the digestive tract; however, complete eradication most likely will require additional efforts with the use of herbal products and prescription drugs. Continued adherence to this program will prevent the repopulation in healthy individuals. Some helpful anti-candida, anti-fungal prescription drugs are Nystatin® (external only), Sporanox® (Itraconazole), and Diflucan® (Fluconazole), but some strains of candida may be resistant. Diflucan® is the recommended drug for clearing fungal infections from the digestive tract and throughout the body. Take at least 150 mg per day for 5 days as an annual treatment. Obvious fungal infections require a higher dose. Your doctor may prescribe a one-day, single-dose treatment only. This treatment may be sufficient for a vaginal yeast infection, but it certainly is questionable for a systemic candida albicans infection.

Herbal products are sometimes helpful, and these can be tested before resorting to prescription drugs. Renew Life CandiGONE® includes many anti-candida herbs such as Arctostaphylos uva ursi (Bearberry) which has been scientifically proven to have antibiotic properties. As a second alternative, take olive leaf extract. Begin with one capsule per day and increase to three. Another choice is oil of oregano enteric coated time-release softgel capsules. Begin with one per day and increase to one per meal. Use oregano with caution as it may cause sudden candida die-off and diarrhea. A fourth alternative is NOW Candida Clear® with Pau D'Arco, oregano oil, black walnut, and caprylic acid.

Chapter 5

Other herbal products are barberry, golden seal, and Oregon grape root.

Diflucan (Fluconazole)
http://www.drugs.com/pro/diflucan.html

Esophageal candidiasis

"The recommended dosage of Diflucan for esophageal candidiasis is 200 mg on the first day, followed by 100 mg once daily. Doses up to 400 mg/day may be used, based on medical judgment of the patient's response to therapy. Patients with esophageal candidiasis should be treated for a minimum of three weeks and for at least two weeks following resolution of symptoms."

Systemic Candida infections

"For systemic Candida infections including candidemia, disseminated candidiasis, and pneumonia, optimal therapeutic dosage and duration of therapy have not been established. In open, noncomparative studies of small numbers of patients, doses of up to 400 mg daily have been used."

"Systemic candidiasis: 400 mg orally or IV once daily."

Your doctor can test your stool specimen for the presence of pathogens, such as yeast, fungus, parasites, and pathogenic bacteria by sending it to Genova Diagnostics®.

"The "Organ Acid Test" evaluates all of the well-defined inborn errors of metabolism that can be detected with this technology (called GC/MS) such as PKU, maple-syrup urine disease, and many others. In addition, we check for

Achieving Remission In Autoimmune Diseases

many other abnormalities such as vitamin deficiencies and abnormal metabolism of catecholamines, dopamine, and serotonin. They currently quantitate 62 substances but also evaluate other substances that are not quantitated. Some of the other biochemical abnormalities common in autism include elevated uracil and elevated glutaric acid. Prescription drugs are recommended for the treatment of any pathogens the tests discover. However, the diet presented here, along with the probiotics, vitamins, and minerals should be started immediately before resorting to stool testing and drugs."

The Organ Acid Test is recommended for:

AD(H)D	Chronic Fatigue Syndrome	Endometriosis
AIDS	Colitis/Crohn's Disease	Fibromyalgia
Alzheimer's Disease	Depression	Multiple Sclerosis or MS-like symptoms
Asperger Syndrome	Bowel Disorders	Psoriasis
Autism & PDD	Down's Syndrome	Recurrent Infections and many more conditions

People with the experience and training in self-administering drugs can purchase drugs for personal use online without a prescription on the following International websites. The drugs must be approved by the Food and Drug Administration (FDA) in the United States. The FDA does not prevent importation when the drugs are for personal use and do not appear to present a serious risk. Individuals also travel to foreign countries for medical treatment where drugs are prescribed by a doctor. The

Chapter 5

patients must continue to receive their drugs upon their return to the United States. Some foreign Internet companies provide a medical service by which a doctor can issue a prescription. The FDA can refuse importation of prescription and non prescription drugs, and the US Customs Service may return the shipment to the supplier. Previous orders placed with the following pharmacies have been very satisfactory:

>EDrugNet.com – United Kingdom.
>EDrugNet.co.uk – United Kingdom.
>MedicinesMexico.com - Mexico.
>Medrx-one.com - Mexico.
>InternationalDrugMart.com - India.
>UnitedPharmacies.com - New Zealand.

The following quotation from a USFDA manual addresses the topic of importation of prescription drugs for personal use. The law does not allow this practice, but the system in the United States has been overwhelmed by the volume of shipments. Some members of the US House of Representatives and Senate have attempted to change the law without success. This is reminiscent of the 55-mph speed limit on the highways in the 1970s and 1980s during which the average citizen refused to obey the unreasonable law. The pharmaceutical companies are fighting hard to keep people from buying lower-priced prescription drugs from foreign manufacturers. Health insurance companies should be pleased because people pay for the drugs themselves. The importation reduces government health-care costs, but some bureaucrats go ballistic when they lose control over the citizens.

FDA - Regulatory Procedures Manual March 2008, Chapter 9 "Coverage of Personal Importations". Quotation as follows:
http://www.fda.gov/ICECI/ComplianceManuals/RegulatoryProceduresManual/default.htm

Achieving Remission In Autoimmune Diseases

"Coverage of Personal Importations"

"PURPOSE

To provide guidance for the coverage of personal-use quantities of FDA-regulated imported products in baggage and mail and to gain the greatest degree of public protection with allocated resources.

BACKGROUND

Because the amount of merchandise imported into the United States in personal shipments is normally small, both in size and value, comprehensive coverage of these imports is normally not justified. This guidance clarifies how FDA may best protect consumers with a reasonable expenditure of resources.

There has always been a market in the United States for some foreign made products that are not available domestically. For example, individuals of differing ethnic backgrounds sometimes prefer products from their homeland or products labeled in their native language to products available in the United States. Other individuals seek medical treatments that are not available in this country. Drugs are sometimes mailed to this country in response to a prescription-like order to allow continuation of a therapy initiated abroad. With increasing international travel and world trade, we can anticipate that more people will purchase products abroad that may not be approved, may be health frauds or may be otherwise not legal for sale in the United States.

In addition, FDA must be alert to foreign and domestic businesses that promote or ship unapproved, fraudulent or otherwise illegal medical treatments into the United States or who encourage persons to order these products. Such treatments may be promoted to individuals who believe that treatments available abroad will be effective in the treatment of serious conditions such as AIDS or cancer. Because some countries do not regulate or restrict the exportation of products, people who mail order from these businesses may not be afforded the protection of either foreign or U.S. laws. In view of the potential scale of such operations, FDA has focused its enforcement resources more on products that are shipped commercially, including small shipments solicited by mail-order promotions, and less on those products that are personally carried, shipped by a personal non-commercial representative of a consignee, or shipped from foreign medical facility where a person has undergone treatment.

PERSONAL BAGGAGE

Chapter 5

FDA personnel are not to examine personal baggage. This responsibility rests with the CBP. It is expected that a CBP officer will notify their local FDA district office when he or she has detected a shipment of an FDA-regulated article intended for commercial distribution (see GENERAL GUIDANCE below) an article that FDA has specifically requested be detained, or an FDA regulated article that appears to represent a health fraud or an unknown risk to health.

When items in personal baggage are brought to FDA's attention, the district office should use its discretion, on a case-by-case basis, in accordance with the guidance provided under GENERAL GUIDANCE below, in deciding whether to request a sample, detain the article, or take other appropriate action.

MAIL SHIPMENTS

FDA personnel are responsible for monitoring mail importations. It is expected that a CBP officer from the CBP Mail Division will examine a parcel and will set it aside if it appears to contain a drug, biologic, or device, an article that FDA has specifically requested be held, or an FDA-regulated article that appears to represent a health fraud or unknown risk to health. FDA should audit those parcels set aside by CBP in accordance with the guidance provided under GENERAL GUIDANCE below, using the following procedures:

Prepare a Collection Report for each parcel sampled. Generally, a physical sample is not required on mail importations because a documentary sample (for example, labeling, labels and inserts) will be sufficient for most regulatory purposes. If a physical sample is needed, collect only the minimum necessary for analysis by the laboratory. The remaining portion should not be removed from the custody of the CBP Mail Division. Importations detained in accordance with this guidance should be held by CBP until they are either released or refused entry. Attached as guidance are two specimen letters that may be sent with the Notice of Detention and Hearing when a parcel is detained. (See Exhibit 9-3 for use in general mail importations and Exhibit 9-4 for use in unapproved drug or device mail importations).

On occasion, products detained by FDA will be mixed with non-FDA-regulated products. When we refuse admission of the FDA-regulated portion, any request for the release of the non-FDA-regulated portion should be referred to the CBP Mail Division with a Notice of Refusal of Admission covering the detained article. Final disposition of all merchandise, including the destruction of detained merchandise, is the responsibility of CBP.

Achieving Remission In Autoimmune Diseases

GENERAL GUIDANCE

The statements in this chapter are intended only to provide operating guidance for FDA personnel and are not intended to create or confer any rights, privileges, or benefits on or for any private person.

FDA personnel may use their discretion to allow entry of shipments of violative FDA regulated products when the quantity and purpose are clearly for personal use, and the product does not present an unreasonable risk to the user. Even though all products that appear to be in violation of statutes administered by FDA are subject to refusal, FDA personnel may use their discretion to examine the background, risk, and purpose of the product before making a final decision. Although FDA may use discretion to allow admission of certain violative items, this should *not* be interpreted as a license to individuals to bring in such shipments."

Are Internet Drugs Safe?

Before purchasing drugs on the Internet, it is best to review the manufacturers' authenticity and standards by looking on their websites prior to placing orders. When the drugs are received, they should be sealed in special bubble-wrap packages with the manufacturer's name, drug name, and dosage clearly printed. The complexity of the package and the printing indicate that the drugs are genuine. Tablets or capsules that are packaged loose in a plain bottle or simple box are totally unacceptable. You can expect the package to have been opened by US Postal Service employees for inspection.

Some of the suppliers above ship high-quality generic and name-brand drugs from multibillion dollar international pharmaceutical companies, such as Ranbaxy Laboratories® (India); GlaxoKlineSmith® (United Kingdom); CIPLA, Limited® (India); Sun Pharmaceuticals Industries, LTD.® (India); Surya Pharma® (India); Unichem Laboratories Limited® (India); Mega Fine Pharma® (India); Ind-Swift Limited® (India); Dr. Reddy's Pharmaceutical Company® (worldwide); and Stada Arzneimittel

Chapter 5

AG® (Germany). Avoid all drugs made in China. Ask your pharmacist if your current prescription drugs are made in China.

Are Prescription Drugs Bad for Us?

Most people resist taking prescription drugs until they are faced with the choice between taking a drug or certain death. Most doctors hesitate to prescribe antibiotics based on the false logic that drugs create drug-resistant bacteria. Many doctors hesitate to prescribe an antibiotic when they suspect the patient has a common food poisoning based on the false logic that the patient's immune system is sufficient to make him better within a few days. The doctor may also fail to test for the pathogen which could later be found to be deadly. A common e-Coli bacterial infection can cause permanent liver damage and death. Food poisoning causes 4,000 deaths in the United States every year. Most food poisoning comes from salad bars, not from meat.

Some prescription drugs have been found to cause death, and the USFDA has pulled many off the market. Other drugs have serious negative side effects, but the bad drugs should not be used as a standard to label all drugs as unacceptable. Many drugs are awesome. For example, one day for lunch I ate some raw green pepper that was contaminated with pathogenic bacteria. I began to experience typical symptoms of food poisoning within 24 hours, so I immediately took one 500 mg Cipro® (Ciprofloxacin hydrochloride) tablet. Some doctors avoid prescribing Ciprofloxicin because of negative side effects; however, the food poisoning symptoms soon dissipated, and I ate a full prime rib dinner without any recurrence of the symptoms. I continued to take the drug for several more days. Placing a blanket rejection label on all prescription drugs can be dangerously unhealthy. Keflex® (cephalexin) has now become my drug of choice for effectively treating food poisoning.

Achieving Remission In Autoimmune Diseases

Are Generic Prescription Drugs Safe?

Generic drugs purchased from your local drugstore may not be as safe as Internet drugs even though you purchased them with a doctor's prescription. The drugs are handled in bulk rather than safety bubble wrap, and in some cases, you may not know the origin of manufacture. They may have come from China where impurities in food, supplements, and drugs are common. They could contain heavy metals, carcinogens, or toxic trace chemicals. Don't assume you are getting the best quality drugs simply because you purchased them with a doctor's prescription from your local pharmacy on your health insurance.

Laboratory Tests You Can Obtain Yourself

If you don't have health insurance, the best approach is to purchase the lab tests online at a discount through HealthCheckUSA® and take your order to the LabCorp® office. This procedure is much less expensive than simply walking into the LabCorp® office.

> HealthCheckUSA®
> Blood and Urine Tests You Can Order Yourself Online
> http://www.healthcheckusa.com/
>
> Laboratory Corporation of America® (LabCorp)
> https://www.labcorp.com/wps/portal/

The best test package is the *Men's Profile* or *Women's Profile*, either of which can be performed at any local LabCorp® facility. Call the laboratory to see what the requirements are, but an appointment is generally not necessary. Simply find the nearest office and walk in with your order.

Chapter 5

Leady Gut Syndrome

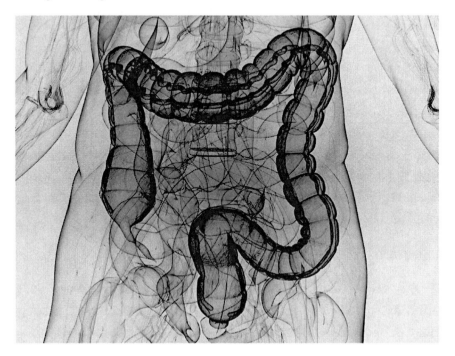

Leaky Gut Syndrome Causes Autoimmune Diseases

The excessive intake of carbohydrates in combination with a protein and fat deficiency causes a condition known as *leaky gut syndrome* in which vegetable protein molecules from grains, corn, soy, and legumes pass through the intestinal barrier and enter the bloodstream. Other food molecules and parts from dead bacteria can also leak into the bloodstream. The body's immune system sees these molecules as invaders and begins to attack them. The immune system attacks the intestinal tract and other areas of the body that contain proteins similar to those found in the vegetable products.

Large doses of antibiotics and non-steroidal anti-inflammatory drugs (NSAIDS such as ibuprofen) are major contributing factors. Antibiotics kill off the good bacteria that break down toxic

Achieving Remission In Autoimmune Diseases

molecules and protect the intestinal tract, thereby allowing the overgrowth of candida yeast. When antibiotics are taken, it is also necessary to take an anti-candida drug and strictly follow the diet presented here. Antibiotics, however, are not the primary cause for bowel diseases as often theorized. Dogs and cats disprove the antibiotic theory. The cheaper pet foods contain corn syrup, corn, soybean, wheat, and other grains as their primary ingredients. Bowel diseases, arthritis, diabetes, and obesity have become epidemic among household pets because they are fed grains instead of meat products. Dogs and cats are strictly carnivorous and should not eat grains. They get bowel diseases even though they have never taken any antibiotics or other drugs. Humans are also very highly carnivorous.

Surgery and drugs will not cure IBD.

Cutting out parts of your colon will not stop inflammatory bowel diseases. They will only spread to another area. Destroying your immune system with prescription drugs will not stop IBD. The only hope is to control the autoimmune response with the diet program presented below in order to heal the digestive system.

Leaky gut syndrome is always associated with autoimmune diseases—a condition in which the immune system makes antibodies against the body's own tissue. Diseases in this category include lupus, alopecia areata, psoriasis, rheumatoid arthritis, multiple sclerosis, fibromyalgia, chronic fatigue syndrome, vitiligo, thyroiditis, vasculitis, Crohn's disease, ulcerative colitis, hives, Sjogren's disease, and Raynaud's disease. Symptoms can include gastrointestinal reflux, pains, diarrhea, cramps, constipation, flatulence, bloating, and bleeding from the rectum. Diarrhea and constipation may at times cycle from one to the other. Unexplained weight loss and the inability to regain

Chapter 5

weight are very common, but weight gain to the point of obesity is common with these diseases and the high-carbohydrate diet.

The protein molecules that make up the delicate lining of the intestines are replaced every three days—the shortest life of any structure in the body. Amino acids must be available for the body to reconstruct this intestinal lining, and glutamine is one of the most important amino acids for a healthy digestive tract. Because the intestines are the only organ in the body that uses glutamine as its primary source of energy, it is very important to eat a high-quality animal protein with every meal and supplement with glutamine. Betaine hydrochloride (HCL) is very effective against candida and is required for protein digestion. Candida thrives in an over-alkaline body-chemistry environment. I urge you to read *Why Stomach Acid is Good for You* by Jonathan V. Wright, M.D., and Lane Lenard, Ph.D.

Taking whey protein powder is highly recommended because it provides amino acids necessary for building bone collagen and healing the intestinal tract, immune system, and nervous system. This is more important in a person with IBD because the digestion may be poor. Do not substitute casein, soy, or egg protein powders because they are not equivalent. Whey protein powder is not a substitute for eating meat, but it can be added to foods or taken between meals with drinks. Whey protein powder is the best choice because it has all of the isolated amino acids as well as branched-chain amino acids (BCAA) and quadra-peptides—short protein chains containing four amino acids.

This combination of amino acids has been shown to provide the following healing properties:

- Provides pain-killing effects by healing the nervous system.

Achieving Remission In Autoimmune Diseases

- Allows body-building amino acids to be absorbed without requiring digestion.
- Stimulates insulin-like growth factor 1 (IGF-1) which functions similarly to insulin and enhances protein synthesis and healing.
- Fights infections by stimulating the immune system. All immune cells are made from poly-peptides of amino acids.
- Provides bone growth of protein collagen and strengthens bones. Poor digestion has been shown to cause osteoporosis and degenerative bone disease.
- Provides all of the amino acids required to heal and grow ligaments, tendons, joints, muscles, intestinal tract, heart muscle, and all other organs of the body.
- Prevents hypoglycemia (low blood sugar) symptoms in people with hypoglycemia or diabetes.

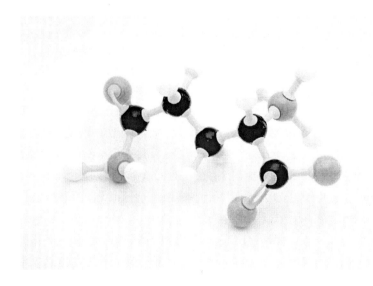

**Model of the Glutamine Amino Acid Molecule
Professional Bodybuilders Take Extra Glutamine**

Chapter 5

Essential Fatty Acids

The three most nutritionally-important omega-3 fatty acids are alpha-linolenic fatty acid (ALA), eicosapentaenoic acid (EPA), and docosahexaenoic fatty acid (DHA). These are classified as *essential* because the body cannot make them from other fats, and deficiencies lead to many diseases. Flax seed oil contains alpha-linolenic fatty acid (ALA) but does not contain EPA or DHA fatty acids. Deficiencies result because some people cannot convert ALA into EPA or DHA. Therefore, flax seed oil is not a good source for the essential omega-3 fatty acids. The extensive campaign to promote flax seed oil is driven by vegetarians and flax seed oil salesmen. Cod liver oil is the best choice because all three of these essential fatty acids are directly available to the body; it does not have to convert ALA to EPA and DHA. Cod liver oil naturally contains vitamins A and D that are essential for bone formation and cancer prevention, but it does not contain unhealthy omega-6 vegetable fatty acids as found in flax seed oil. Do not eat flax seed oil or put ground flax seed in your food.

Linoleic fatty acid (LA) and arachidonic fatty acid (ARA) are the two essential omega-6 linoleic fatty acids. In reality, only ARA is essential, but LA has also been listed because the body can make ARA from LA by a process known as *biosynthesis*. One or the other is needed—not both. These fatty acids are essential since the body is unable to make them from other fats and because they play a fundamental role in several physiological functions. A deficiency leads to many diseases. Therefore, we must be sure our diet contains sufficient amounts of either linoleic acid or arachidonic acid. The body easily changes linoleic fatty acid—found mainly in grain, nut, and seed oils—into arachidonic fatty acid, but the process can create an unhealthy oversupply of ARA. Vegetarian diets result in an unhealthy oversupply of ARA because an oversupply of linoleic acid is found in all vegetable oils. ARA is pro-inflammatory when the intake is excessive. Arachidonic fatty acid is the essential omega-6 fatty acid that is found in a healthy but limited amount in meat. Although most dietary experts teach that omega-6 polyunsaturated fats from vegetable oils are healthy, they definitely are not. The following links are three of many examples that expose this misconception. Vegetable oils should not be called *healthy* when research shows they cause cancer, heart disease, asthma, and other autoimmune diseases.

"Dietary polyunsaturated fatty acids and inflammatory mediator production"
http://www.ncbi.nlm.nih.gov/pubmed/10617994?dopt=Abstract

> "The pro-inflammatory eicosanoids prostaglandin E(2) (PGE(2)) and leukotriene B(4) (LTB(4)) are derived from the n-6 fatty acid arachidonic acid (AA), which is maintained at high cellular concentrations by the high n-6 and low n-3 polyunsaturated fatty acid content of the modern Western diet." Omega-6 fatty acids should be avoided. They are found in vegetable, seed, and nut oils such as safflower oil, corn oil, soybean oil, peanut oil, and others."

"Healthy fat link to bowel disease"
BBC New Health - July 24, 2009.

> "A high intake of polyunsaturated fat in the diet, while good for the heart, may lead to inflammatory bowel disease, say researchers. Experts believe a high intake of linoleic acid, found in foods like "healthy" margarines, may be implicated in a third of ulcerative colitis cases."

Chapter 5

"Margarine 'may increase asthma risk"
BBC New Health - July 19, 2001.

> "A diet high in polyunsaturated fats - found in many margarines and vegetable oils - may double a child's chances of having asthma, according to researchers."

The edible oil industry produces omega-6 polyunsaturated vegetable, seed, and grain oils, such as corn, soybean, safflower, sunflower, peanut, and cottonseed oils. These oils are thought to be among the leading causes of heart disease and cancer—both of which increase in concert with increases of omega-6 fatty acids in the diet. Omega-6 fatty acids are pro-inflammatory and have been proven to cause or contribute to a long list of autoimmune diseases, such as rheumatoid arthritis, lupus, multiple sclerosis, Crohn's disease, fibromyalgia, irritable bowel syndrome, inflammatory bowel disease, and many others.

Healthy Fatty Acid Foods

Animal Fats	Arachidonic Acid	Saturated Fats
Fish Oils	EPA Omega-3	DHA Omega-3
Butter	Coconut Oil	Monounsaturated

Unhealthy Fatty Acid Foods

Grain Oils	Seed Oils	Most Nut Oils
Vegetable Oils	Omega-6 Fats	Polyunsaturated
Trans Fatty Acids	Soybean Oil	Corn Oil

Hyperglycemia (high levels of blood glucose resulting from glucose intolerance) in the presence of hyperinsulinemia (high levels of blood insulin caused by insulin resistance) causes polysaccharide chains to attach to serum proteins, artery

Achieving Remission In Autoimmune Diseases

proteins, serum hemoglobin, and LDL molecules in a process called *glycation*. The glycated molecules and rancid omega-6 fatty acids form a fatty deposit called *atherosclerotic plaque*. This is the solid scientific proof that glucose from dietary carbohydrates causes heart disease.

Salmon Steak Provides All of the Essential Fatty Acids and Amino Acids

Fruit and Vegetable Salads Provide None

Chapter 5

Pathogenic Bacteria Infections in the Intestinal Tract

Most of the digestion and absorption of nutrients occurs in the small intestine. Its long length alone provides a large surface area for digestion and absorption, and that area is further increased by mucosa villi on the surface of the interior intestinal wall. The mucosa—inside surface—and the submucosa allow the small intestine to complete the process of digestion and absorption. The mucosa forms a series of villi, or small projections, only 0.5 to 1.0 mm high which give the intestinal mucosa a velvety appearance. These villi, the primary structures responsible for the digestion and absorption of food, become damaged from a diet of excessive carbohydrates and the deficiency in proteins and fats. A protein deficiency may be the key because the villi have a short life and are continually being replaced. When there is a deficiency of one or more of the essential amino acids, this replacement may not occur. Digestion of the carbohydrates in the small intestine then becomes incomplete, and undigested carbohydrates pass to the colon—large intestine—where they don't belong. This presence of carbohydrates in the colon encourages the growth of pathogenic bacteria—disease causing bacteria, such as Giardia, Proteus Vulgaris or Clostridium difficile—and opportunistic pathogenic fungi. There are several possible therapeutic alternatives for treating intestinal pathogenic bacteria. Clostridium difficile can be deadly and is the most difficult to kill. It requires Flagyl® (metronidazole) or Vancomycin. Flagyl® is an awesome drug. It also kills amoeba (protozoa) in addition to the resistant bacteria, Clostridium difficile (C. diff). Take 7.5 mg of Flagyl per 2.2 pounds (1.0 kg) of body weight three times a day. A person weighing 150 pounds would take 500 mg three times a day for 10 days or as recommended by his doctor. Flagyl ER® (extended release) is a good choice at 750 mg per day for 10 days.

Achieving Remission In Autoimmune Diseases

Intestinal floras (bacteria) are unevenly distributed in the digestive tract where they increase as food moves through the small intestines and into the colon. The colon contains a mass of bacteria, and the stool contains a very high percentage of live and dead bacteria. The stool is approximately 75% water and 25% solids. Live and dead bacteria comprise as much as 50% of the solids. The remaining 50% consists of residual products of digestion. The colon contains thousands of different bacteria. Some are essential; some are neutral; some are unhealthy; some are pathogenic; and in rare cases, the pathogens are deadly. These pathogenic bacteria in the colon are commonly associated with most bowel diseases because they produce toxins that irritate the digestive system and cause symptomatic reactions in the body. The intestinal tract takes on the characteristics of a hazardous waste dump.

Bad bacteria and candida yeast can overpower and destroy the healthy intestinal floras. Treating pathogenic bacteria with drugs is only a temporary solution because the opportunity exists for their reestablishment. Proper diet combined with the reestablishment of good bacterial floras is essential. The bacteria must be starved to death by removing their primary food sources from the diet—carbohydrates and fiber. Removing these undesirable bacteria by drug treatment will not be successful if carbohydrates and fiber continue to be consumed. They will simply return after drug treatment ends because they are opportunistic.

When people have their fecal matter tested and find that one or more pathogenic bacteria are present, they assume they must have somehow contracted the bacteria. They also assume that this is the reason for their inflammatory bowel disease. Both of these assumptions are wrong. People are routinely exposed to

Chapter 5

these bacteria which are suppressed by a healthy digestive system. The plethora of carbohydrates and fiber in the diet allows the opportunistic pathogenic bacteria to proliferate and make the IBD worse. Your doctor can send your stool specimen to Genova Diagnostics® to test for the presence of pathogens.

The abundance of pathogenic bacteria in the intestines produced by the consumption of carbohydrate foods, such as whole grains, fruit, and starchy vegetables, proliferates in the appendix where they set off chronic inflammation. As a result, appendectomies are very common in people with IBD.

Exposure to Pathogens from Household Pets, Farm Animals, and Environmental Hazards

Pets in the home present us with serious health risks from exposure to pathogens. Cats scratch in their litter pans and track feces all over the house where the occupants come in contact with viruses, parasites, staphylococcus, streptococcus, and other bacteria. Dogs lick everywhere and everyone. Farm animals present a pathogenic occupational hazard. The dust and pathogens fly in the air when chicken are startled in the chicken house. Pathogens fly off trucks that are transporting chickens, cows, and pigs. The neighborhood garbage collection truck is spewing a plume of pathogens as you drive behind it in traffic. I refuse to follow these trucks and either stop or change my route.

Colon Cleanup of Undesirables and Pathogenic Bacteria

A routine or annual colon cleanup with an antibiotic is recommended to clear the colon of bad bacteria that creep into the digestive tract. This is a healthy procedure when done correctly. Even those who show no sign of pathogenic bacteria in a stool analysis should take a treatment of antibiotics to eliminate these bacteria that could be continually disturbing

Achieving Remission In Autoimmune Diseases

digestion. Flagyl® (metronidazole) is the recommended prescription drug of choice for a broad spectrum of pathogenic and deadly bacteria and has been shown to be one of the drugs of choice for Clostridium difficile (C. diff) and Escherichia coli (E. coli) bacterial infections in the colon. An annual treatment should be considered to clear the digestive tract of pathogenic infection. Take 7.5 mg of Flagyl® per 2.2 pounds (1.0 kg) of body weight three times a day. A person weighing 150 pounds would take 500 mg three times a day for 10 days or as recommended by his doctor. Flagyl ER® (metronidazole extended release) is a good choice at 750 mg per day for 10 days.

Keflex® (cephalexin) is another generally well-tolerated drug to kill Escherichia coli bacteria as well as other pathogenic bacteria in the digestive tract, but it is not effective against C. diff. Take one 500 mg capsule three times a day for 5 days and reduce to one 500 mg capsule one time a day for 10 days or as recommended by your doctor.

Ciprofloxacin and ofloxacin have the undesirable side effect of possible tendon weakness or damage, and they are not effective against Clostridium difficile infections in the colon. A visible side effect is marks or grooves in fingernails as they grow out. Do not take ciprofloxacin or ofloxacin in high doses or for an extended period of time because it can cause severely ruptured tendons. As an alternative, begin by taking 250 mg of ciprofloxacin twice a day with meals (breakfast and dinner) for the first two days, followed by 250 mg once a day with breakfast for five more days or as recommended by your doctor. As another alternative, take one 300 mg ofloxacin tablet with breakfast for five days. Doxycycline may have fewer side effects but is not as effective.

Failure of the diet presented here is a sure sign that the colon has an infestation of undesirable bacteria. Another sign is bad breath.

Chapter 5

A stinging sensation upon urination is a sign the infection has spread to the bladder and kidneys.

Probiotics should also be taken at the same time in order to replace the undesirable floras with these better choices. Take one each of the following two probiotics three times a day with meals for 10 days followed by one each per day with breakfast on a continuous basis thereafter. The daily supply of these good probiotics helps to prevent the growth of undesirable bacteria.

> **Multidophilus®** is a lactic flora by Solaray®.
> Buy locally from a refrigerated display case and keep refrigerated. Do not leave it in a hot car.
>
> **Lactobacillus Sporogenes®** is a probiotic by Thorne Research®.
> It is difficult to find locally but can be shipped without refrigeration. Refrigerate after opening.
>
> ***Benefits of Lactobacillus Sporogenes as a probiotic***
> http://www.lactospore.com/benefit.htm

Colon Cleanup of Yeast and Fungal Infections

Antibiotics have been criticized for making digestion worse. This criticism is correct if the fungi are left to proliferate untreated, but a blanket criticism of antibiotic is unwarranted. People with IBS or IBD generally have colons that are severely infected with pathogenic bacteria and fungi that are in competition for the food supply—carbohydrates and fiber. Antibiotics will kill the pathogenic bacteria as well as the good intestinal floras. The yeasts and fungi remain to flourish. The results make it appear that the antibiotic made the IBS or IBD symptoms worse.

A routine or annual colon cleanup with an antifungal drug is recommended at the same time as the antibiotic cleanup. This balanced approach keeps the offensive organisms from growing out of control. Diflucan® is the recommended drug for clearing fungal infections from the digestive tract and throughout the body. Take at least 150 mg per day for 5 days as an annual treatment or as recommended by your doctor. Obvious fungal infections require a higher dose.

Carbohydrates Cause Leaky Gut Syndrome and Autoimmune diseases

Carbohydrates cause an increase in insulin which promotes the release of epinephrine (adrenalin). High levels of epinephrine set off a collection of physiological reactions called the *fight-or-flight responses*. When these responses occur in the gastrointestinal tract, they cause muscular movements and digestive secretions to slow down or stop and the result is constipation. Most gastroenterologists erroneously recommend a high-fiber diet to combat constipation from the carbohydrates, but the correct action should be to stop eating carbohydrates. The fiber can be digested by pathogenic bacteria and will encourage their growth. These bacteria stimulate intestinal secretions which produce diarrhea. This is why so many people with intestinal diseases find no relief from eating fiber or taking fiber supplements. High-fiber foods are always high-carbohydrate foods. The body responds by pulling out the water to prevent offensive carbohydrates from entering the colon—constipation. The body can also reverse the process by pouring lots of water into the colon in order to flush out the offensive carbohydrates—diarrhea. When a sudden switch from one to the other and back occurs, the vicious cycle is underway.

The low-carbohydrate, low-fiber diet recommended here has been shown to stop this vicious cycle. This high-protein, high-fat diet reduces insulin and adrenalin responses; it promotes

Chapter 5

healing, a healthy digestive system, and normal bowel movements. These improvements occur in a matter of weeks, or days in most cases, simply by eating a diet typical of many primitive cultures throughout the centuries. This was the natural diet for humans prior to the cultivation of grains and the production of factory foods. In order to fully appreciate this diet plan, you must overcome the current nutritional myths which are promoted by food manufacturing companies.

Fiber and carbohydrates turn the colon into a fermenting hazardous waste dump.

Some people have a defective gene that makes them carbohydrate intolerant, i.e., unable to absorb glucose and galactose in the small intestine. Glucose-Galactose Malabsorption is inherited as a recessive genetic trait. The sodium-glucose cotransporter [SGLT1] located on the long arm of chromosome 22 (22q13.1) is the gene responsible for this defective disorder. The unabsorbed glucose and galactose continue through the intestinal tract to the colon where they promote the growth of pathogenic bacteria, fungi, and yeasts.

When infants are fed the standard high-carbohydrate diet, their journey to insulin resistance begins. This has been proven to be the root cause of childhood and adult obesity, diabetes, allergies, and heart disease. On the other hand, some children are exceedingly thin and fail to thrive. Instead of receiving the correct diagnosis of Crohn's disease, they are often misdiagnosed as anorexic. High-carbohydrate infant foods are setting children up for a lifetime of health problems.

Achieving Remission In Autoimmune Diseases

"Gastric Ulcers, Peptic Ulcers, and Helicobacter Pylori Bacteria Infections"

BBC Health News July 31, 2008.

"Helicobacter pylori, proved to be the cause of most stomach ulcers, has also been linked with stomach cancer. In a study of 550 people who had stomach cancer surgery, antibiotics which killed the bug cut the risk of a second cancer developing by two-thirds."

A *peptic ulcer* is a hole that forms in the lining of the stomach, upper intestine, or esophagus. An ulcer in the mucous lining of the stomach is called a *gastric ulcer*. One in the uppermost part of the small intestine (duodenum) is called a *duodenal ulcer*.

Mastica—the resinous gum of a species of Greek pistachio tree, Pistacia lentiscus—has been used in Greece for hundreds of years as a remedy for a broad range of gastrointestinal disorders. It has been found to be an effective alternative to pharmaceuticals in the treatment of gastritis, gastro-esophegeal reflux disease (GERD), and many types of intestinal inflammation. It is a safe and effective alternative to antibiotics in the treatment of stomach and duodenal ulcers and the eradication of the bacterium which frequently causes these conditions—Helicobacter pylori (H. pylori). Unlike antibiotics, Mastica does not eradicate the populations of friendly intestinal bacteria that are so crucial to health and well-being.

Source: *New England Journal of Medicine*

"Even low doses of mastic gum -- 1 g per day for two weeks -- can cure peptic ulcers very rapidly, but the mechanism responsible has not been clear. We have found that mastic is active against Helicobacter pylori, which could explain its therapeutic effect in patients with peptic ulcers."

"Mastic is a resinous exudate obtained from the stem and the main leaves of Pistacia lentiscus. It is used as a food ingredient in the Mediterranean region. Clinically, mastic has been effective in the treatment of benign gastric ulcers (1) and duodenal ulcers. (2) In rats, mastic showed cytoprotective and mild antisecretory properties. (3) We assessed the antibacterial properties of mastic against H. pylori."

There are several possible therapeutic antibiotic alternatives for treating Helicobacter pylori. Because this strain is very difficult to eradicate, it must be treated with antibiotics. The diet program presented here should help prevent the repopulation of the bacterium that causes stomach ulcers.

Gastroesophageal Reflux Disease
http://www.intelihealth.com/

"Gastroesophageal reflux disease, commonly called GERD, is a digestive disorder in which the stomach's juices (acid and digestive enzymes) flow backward, or reflux, into the esophagus. The esophagus is the tube that carries food from your mouth to your stomach. The lining of the esophagus can't handle these caustic substances, so the esophagus becomes inflamed. This causes heartburn and other symptoms. If GERD is not treated, it can cause permanent damage to the esophagus."

The lower esophagael sphincter is a valve between the esophogus and the stomach that prevents stomach acid from refluxing into the esophogus. The pyloric sphincter is a valve at the discharge end of the stomach where it empties into the duodenum (the beginning of the small intestines). The lining of the stomach secretes several different gastric products necessary for the healthy digestion of food. These products include hydrochloric acid (HCL) that is used to assist in the digestion of protein in meat and other foods. The pyloric sphincter valve opens when the acid reaches a desirable high level. A low acid level in the stomach causes the pyloric valve to remain closed which allows gases to form. These gases force stomach acid up into the sensitive esophagus, and the result is the burping associated with acid reflux. Acid reflux occurs because acid level is below normal—not because of excessive stomach acid. The lower esophagael sphincter valve can also become damaged which leads to an incurable condition. Acid reflux should not be allowed to continue because it damages the esophagael sphincter valve and the lining of the esophagus.

Drugs to treat acid reflux have been linked to an increase in bacterial infections and bone fractures.

Acid reflux is also caused when the stomach discharges acid at the wrong time—when the stomach is empty—or when it does not discharge enough acid to open the pyloric sphincter valve. Even though young people have much higher levels of stomach acid than older adults, they rarely have GERD. Older people do not have enough stomach acid as is described in *Why Stomach Acid is Good for You* by Jonathan V. Wright, M.D., and Lane Lenard, Ph.D. Taking Betaine HCL with meals may help somewhat, but there is more to it than that. GERD generally disappears within a few days on the low-carbohydrate diet presented below and by supplementing with Betaine HCL. You must eat a low-carbohydrate diet in addition to taking Betaine HCL in order to completely prevent GERD.

Chapter 5

Recent Testimony from a Man with Acid Reflux

"Hi Mr. Rieske,

I am Arnold from Jacksonville Florida. I thank God for leading me to your website. The dietary philosophy that you recommend has helped me with my chronic hyper acidity. I was taking Toms and other antacid for years, but now I no longer need them. I also appreciate the many articles on Biblical truths on your website."

Stomach acid reduces and restricts stomach bacteria, viruses, yeasts, and fungi, although many bacteria and viruses can survive the normal level of acid. Taking antacids or drugs to reduce stomach acid is the wrong approach. Low stomach acid promotes the growth of pathogenic bacteria, viruses, yeasts, and fungi and leads to inflammatory bowel diseases. Never take antacid tablets or drugs to reduce stomach acid.

Parasitic Intestinal Infections

Blastocystis Hominis and Dientamoeba Fragilis are protozoal parasites that can cause constipation and/or diarrhea, mushy stools, stomach pain, bloating, chronic fatigue, and weight loss. The diet program presented here will not eradicate parasites in the digestive tract. Additional efforts with herbal products and prescription drugs will be required. It is necessary to test a stool specimen order to identify the specific parasite. Your doctor can send your specimen to Genova Diagnostics® to test for the presence of pathogens.

Do not fall for the "colon cleanse and detoxification" scam. The colon does not contain a lining of putrid material that looks like "chunks of debris that resemble cooked liver, long black twisted

Achieving Remission In Autoimmune Diseases

rope-like pieces." A colon cleanse or a detox program of harsh herbs, fiber products, fruit, or fruit juices only serves to create more problems that may lead to leaky gut syndrome.

Associated Autoimmune Disease and Disorders

Irritable bowel diseases can predispose one to Alzheimer's disease because the digestive tract becomes acidic. Many doctors and nutritionists erroneously recommend a high-carbohydrate diet containing very little meat and fat as a way to reduce acidity, but that diet is actually the cause of the problem. The acid increases the amounts of aluminum, cadmium, lead, mercury, tin, and zinc that will be absorbed. Aluminum has been linked to Alzheimer's disease along with a deficiency in Vitamin B-12 which may not be absorbed due to the excess acid. The diet I present will return digestive system acidity to normal.

Interstitial Cystitis

> "Interstitial cystitis is a puzzling bladder condition in which the bladder wall becomes irritated or inflamed, causing pain and frequent or painful urination. The symptoms of interstitial cystitis are often similar to the symptoms of a urinary-tract infection. However, in interstitial cystitis, there is no infection, and the symptoms do not respond to antibiotic treatment." InteliHealth.
> http://www.intelihealth.com/

Interstitial cystitis is an autoimmune disease caused by the excessive consumption of carbohydrates, particularly grains. The autoimmune reaction is most likely triggered by leaky gut syndrome. The **Starting Diet** presented here will probably bring prompt relief, but a cure is an impossible goal in autoimmune diseases. The best hope is for remission of the symptoms. This

Chapter 5

diet plan must be continued to prevent recurrence of the disease called a *flare*.

Proctitis

Proctitis is inflammation of the rectum. The flesh can split and form fissures that become infected with bacteria. Healing is difficult because it is an autoimmune reaction that forms the fissure. This diet program in combination with antibiotics should heal the fissures. Infectious viruses can also cause proctitis.

A Fistula is a narrow passage from an abscess to the surface of the skin. Doctors frequently perform surgery on fissures and fistulas, but the autoimmune reaction is most likely to make another attack nearby. A combination of surgery and the nutritional program presented here is required to provide a successful result.

**Illustration of a Healthy Colon
as Seen during a Colonoscopy**

Achieving Remission In Autoimmune Diseases

Starting Diet Overview

Grains, Legumes, Fruit, Starches, Sugars, Milk, Nuts, Fiber, and Vegetable Oils Are Your Enemy

Yes, fiber. That was not a typo. Read on and you will understand why the recommended high-fiber diet is actually bad for intestinal diseases—not the highly-praised cure. The discussion below will concentrate on diet and the digestive system because of the close relationship between all autoimmune diseases and leaky gut syndrome.

Unhealthy High-Fiber Foods

Bran Cereal	Oats	Rice	Soybeans
Legumes	Split Peas	Lentils	Apples
Broccoli	Potatoes	Grapefruit	Pears

Many different diseases in the bowel disease family can occur in any area of the digestive system from the mouth to the anus. The best opportunity for healing or control of these diseases is to eliminate the offending carbohydrates, vegetable oils, fiber, and vegetable protein from the diet and restore the essential proteins and fats from animal products. This means the total elimination of all grains, legumes, fruit, starchy-vegetables, sugars, milk, nuts, seeds, fiber, and vegetable oils from the diet. Carbohydrates and cellulose fiber in the diet end up in the colon where they feed fermentative pathogenic bacteria, yeasts, molds, and fungi that produce alcohol, acetaldehyde, lactic acid, acetic acid, and a host of other toxic chemicals. Intestinal gas—flatulence—is a sure sign that fiber and/or sugars are being fermented. The result can be constipation, diarrhea, bleeding, ulcers, fissures, fistulas, bloating, pains, flatulence, or a burning and itching rectum. The

Chapter 5

continued irritation leads to Crohn's disease, ulcerative colitis, fissures, and the entire list of inflammatory bowel diseases. The vegetarian concept of turning the intestines into a fermentation tube is ridiculous. Avoid all whole grains, brown rice, fruit, starches, and dried beans as they are high in both fiber and complex carbohydrates—a double blow to the digestive system.

Unhealthy High-Carbohydrate Foods

Grains	Cereal	Bagels
Legumes	Fruit	Bread
Pasta	Potatoes	Yams
Sugar	Ice Cream	Candy

Food allergies and autoimmune diseases are caused by partially-digested food crossing the intestinal barrier and entering the bloodstream where immune reactions occur. You must stay on an allergy-free diet to prevent an autoimmune response, but progress can also be wonderfully fast.

This is not a typical low-carbohydrate diet. On page 91 in *Protein Power*, Drs. Michael R. Eades and Mary Dan Eades state, "Aim for 25 grams of fiber each day." My diet aims for zero grams of fiber per day. People with seriously debilitating IBD achieved remission more quickly when the fiber intake was deliberately minimized. On page 67 in *Dr. Atkins' New Diet Revolution*, Dr. Robert C. Atkins states that people should supplement their diets with "one tablespoon of psyllium husks daily" for increased fiber. This supplement is strictly forbidden on my special diet for people who suffer from autoimmune diseases.

Achieving Remission In Autoimmune Diseases

Yikes! What Can I Eat?

The most difficult part of this diet is the mental anguish encountered in trying to overcome the myths and lies from the past that cloud the mind. It is extremely difficult for people to believe that whole grains, bagels, 7-grain muesli cereal, pasta, bread, brown rice, nuts, legumes, milk, yogurt, organic fruit, honey, potatoes, soybeans, and vegetable oils are making them sick. They have always been taught that eating red meat and saturated fat is unhealthy. I have performed years of scientific investigation, monitored the comments of hundreds of IBD sufferers, and received hundreds of emails that support the truths contained in this diet for autoimmune diseases. Intense study of this diet plan is absolutely necessary for you to overcome past erroneous thinking. Chapter 6 will explain why and how we all became brainwashed into believing dietary and medical propaganda that made us sick and will prevent our prompt recovery.

One woman with severe asthma and chronic sinus congestion has tried everything but this nutritional program. She took an antibiotic for the sinus problem but it did not help. She had sinus surgery that I said would not be successful. It was not. She tried acupuncture that I said would not be successful. It was not. She is taking steroids to help keep the asthma under control, but it is marginally successful as many asthmatics can attest. She insists that changing her diet will be the last resort. She will not give up her favorite foods, but these are the very reason her asthma remains in a constant flare. **People are very skeptical that a diet change can help their autoimmune diseases because they have tried other diet programs that did not help.**

Chapter 5

Your First Step
Acknowledge that the foods you have been eating made you sick and be willing to change your diet.

You may eat all meats, fish, fowl, seafood with their natural fats, and a small list of specially selected vegetables. Avoid all prepared meats, such as hamburger, ham, sausage, hot dogs, etc. because they can contain hidden carbohydrates in the form of sugars and starches. Meat labeled 100% ground beef is probably OK. Do not buy any meat, chicken, turkey, or fish that has a flavor-enhancing additive of any kind—period.

The digestion of meat and natural fats does not require the intestinal bacteria needed for digestion of carbohydrates. Stomach acids and enzymes facilitate the digestion of meat proteins into their individual amino acids which pass through the intestinal wall and enter the bloodstream. Those with bowel diseases are able to digest short-chain saturated fats more easily than other fats. The saturated fat is good for those who are underweight and attempting to gain weight; coconut oil is a very good natural short-chain saturated fat for this purpose.

All natural animal fats are healthy, and this diet contains about 70% fat on a calorie basis. Many organizations and professionals are against the intake of fats because they fail to differentiate between the good animal fats and the harmful omega-6 fatty acids.

People with bowel diseases can have amazingly low cholesterol readings because the body is not digesting food, not because their diets are healthy. This diet will raise cholesterol, lower triglycerides, and produce good risk ratios.

Achieving Remission In Autoimmune Diseases

Healthy High-Protein and High-Fat Foods

All Fresh Meat	Beef	Lamb
Pork	Chicken	Turkey
Wild Game	Fish	Seafood
Shell Fish	Hard Cheeses	Coconut Oil
Butter	Organ Meats	Eggs

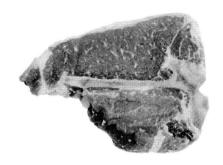

Beef and other red meats are some of the healthiest foods. These meats are naturally loaded with essential vitamins, minerals, enzymes, and a multitude of other dietary nutrients needed for optimum health. Red meat is rich in heme-bound iron, the natural form of essential iron in the body. People who avoid red meat and vegetarians (who avoid all meat) can be iron deficient. The body can also eliminate excess heme iron naturally, unlike elemental iron obtained from supplements that can reach toxic levels. Nutrition books typically underplay or ignore the broad spectrum of nutritional elements in red meat.

This book will prove the healthiest diet for humans is:

70% total fat on a calorie basis
    31% saturated fat
    7% polyunsaturated fat
    25% monounsaturated fat
    7% other fats
27% protein
3% carbohydrates (20 gm of which 3 gm or less is fiber).

Chapter 5

One thing is certain. You don't have constipation or diarrhea because you eat too many Porterhouse steaks. Red meat and saturated animal fats do not cause inflammatory bowel diseases, heart disease, or cancer. Those claims are all myths, lies, and politically-correct nutritional propaganda. People have constipation or diarrhea because their diets consist primarily of high-fiber, high-carbohydrate foods, such as whole grain breads, bagels, cereals, pasta, rice, legumes, fruit, fruit juices, and potatoes. Sugar adds to the problem, although it does not contain fiber.

Supermarket beef steaks will heal your body. Amino acids in meat are the building blocks of life.

The claims that supermarket beef contains harmful antibiotics, hormones, pesticides, toxins, diseases, and bad fats are simply false. These falsehoods have been propagated by vegetarians, grass-fed beef salesmen, and carbohydrate food manufacturing companies.

Antibiotics and hormones given to animals are withheld for a period of time before the animals are slaughtered. This waiting period allows the animals to clear the antibiotics and hormones from their bodies. Supermarket meat from feedlot steers does not contain any harmful antibiotics or hormones. This is one area in which USDA regulations are helping to make our food supply healthier.

Pesticides are sometimes sprayed on farm animals and feedlot steers to rid them of biting and infectious insects. This is extra work and expense, but some producers consider them to be justified. Pesticides that enter the animals through their hides are allowed to clear from their bodies before they are

Achieving Remission In Autoimmune Diseases

slaughtered. Supermarket meat does not contain unhealthy levels of pesticides. Don't believe the lies propagated by the animal rights terrorists.

Attacks against eating red meat have been taken up by the USFDA in their claims that antibiotics given to farm and feedlot animals have created drug resistant bacteria that are a threat to humans. They have also claimed for years that antibiotics given to people are doing the same. These claims are not true. In recent years, hospitals in the United Kingdom have been fighting a serious infestation of the drug-resistant bacterium, Methicillin Resistant Stapholococcus Aureus (MRSA). The same claim was made that antibiotics given to people and animals created MRSA. We now know the MRSA bacterium has been prevalent in the oceans for unknown eons. The USFDA is steering people away from eating meat by promoting fruit, vegetables, and wholegrains. The brainwashing continues.

Beef Half Sides Ready to Ship to Market

Chapter 5

Rieske Autoimmune Diet

Proper Diet Summary

After several years of searching medical, health, and nutritional scientific studies, I developed the IBD healing program presented here. Because many studies reach opposite conclusions, the key is to selectively reject those studies that have a biased agenda. I monitored an Internet IBD message board for three years to obtain actual feedback from one thousand IBD sufferers. From this long-term observation, I identified the list of healing foods and the list of offending foods. Board members identified the **Forbidden Foods** by their severely negative reactions to them. High-carbohydrate foods are always on the **Forbidden Foods** list, but some low-carbohydrate foods are also offensive. Cauliflower, either raw or cooked, is a good example. It causes negative digestive symptoms in most people with IBD. The exact reason is not clear because low-carbohydrate foods are generally less offensive.

The results this diet program produces are much different from those obtained by diets that most medical professionals recommend. People on this diet program have consistently confirmed that their IBD symptoms quickly subside and normal digestion is restored. This diet program is effective for all inflammatory bowel diseases as well as many other autoimmune diseases, such as asthma, rheumatoid arthritis, multiple sclerosis, and lupus. The information and references contained in this book will start you on the proper diet for healing and health preservation. Fiber is not the IBD sufferer's friend.

Give this program a week and it will give you a life.

Achieving Remission In Autoimmune Diseases

High-Protein, High-Fat, and Low-Carbohydrate Diet

High-Protein, Low-Carb Breakfast
Two fried eggs topped with Swiss cheese. Avocado, tomato juice with L-glutamine amino acid, whey protein powder, and colloidal minerals. Take Carlson's® Lemon Flavored Cod Liver Oil and supplements.

High-Protein, Low-Carb Lunch
Left-over baked chicken with stewed combination of eggplant and tomatoes. Reverse osmosis water, vitamins, minerals, and supplements.

High-Protein, Low-Carb Dinner
7-bone beef roast cooked in a crock pot with diced tomatoes. Reverse osmosis drinking water, vitamins, minerals, and supplements.

High-Protein, Low-Carb Breakfast
100% ground beef fried in coconut oil topped with Swiss cheese. Add fried mushrooms and boiled asparagus. Reverse osmosis water, potassium chloride salt substitute, vitamins, minerals, and supplements.

High-Protein, Low-Carb Lunch
Gas-grilled lamb riblets. Cooked red peppers and tomatoes. Reverse osmosis water, cod liver oil, potassium chloride salt substitute, vitamins, minerals, and supplements.

High-Protein, Low-Carb Dinner
Baked farm-raised salmon fillet with boiled asparagus. Enjoy real butter on cooked vegetables. Reverse osmosis water, potassium chloride salt substitute, vitamins, minerals, and supplements.

Chapter 5

Warning!
Please read the Medical Disclaimer above.

The program presented here consists of an abundant selection of fresh foods that have been shown to cause a low allergic response. The vitamins and supplements listed are commonly available in local vitamin stores or on the Internet. This program is not intended to replace the treatment recommended by your physician. You should discuss this with a physician who is versed in the benefits of the low-carbohydrate diet before beginning this diet and supplement program.

Initial Low-Carb Symptoms Are More Severe in Diabetics

Going from a standard high-carb diet to the low-carb diet can initially cause several negative symptoms depending on other health problems. Typical temporary symptoms are:

- Slight headache. Simply ignore it.
- Ketosis is not harmful at all, but it does cause a funny taste in the mouth. This will only last for a few weeks at most.
- Slight constipation or sluggish bowel. This improves to normal in a few weeks. Digestion on low-carb is perfect with very little gas. Skipping days with no bowel movement is normal and expected.
- Leg muscle weakness at the beginning of a hike. Simply keep going. The strength will return in a few minutes, and the legs will be stronger than ever as you continue on the hike.

Carbohydrates are HIGHLY addictive, but people don't believe this until they try going without them. This is the main reason people give up on the low-carb diet—especially those who are prone to addictive behaviors and have weak self control. They

Achieving Remission In Autoimmune Diseases

make up all kinds of excuses for quitting low-carb, but the truth is they can't overcome their addictions to carbohydrates. Addiction to high-sugar fruit is a serious problem among adults and children. A study of the shopping carts of adults and the eating habits of children easily proves this unhealthy addiction exists in all age groups They would rather be sick than give up the addiction.

Starting Diet

- **Full Nutrition.** This is a high-protein, very high-fat diet. This IBD **Starting Diet** helps all autoimmune diseases, diabetes, heart disease, fibromyalgia, chronic fatigue syndrome, polycystic ovarian syndrome (PCOS), osteoporosis, and cancer. The diet has an abundance of all of the nutritional essentials and eliminates all the offending foods, such as sugar, fructose, grains, fiber, starches, other carbohydrates, trans fats, and omega-6 vegetable, nut, and seed oils. Remember, eating only one of the offending foods will cause severe symptoms; it will seem as if the diet is not helping at all. Therefore, it is extremely important that the diet program be followed to the letter. The extra vitamins, minerals, and supplements are very important for a surcharge of essential nutrients. These help to overcome the deficiency in vitamin and mineral absorption that is common in IBS and IBD due to intestinal damage. Adhere to the vitamin, mineral, and supplement program that follows. Adding colloidal trace minerals to the list of other minerals is extremely important. Symptoms of vitamin K deficiency include easy bruisability, epistaxis, gastrointestinal bleeding, menorrhagia, and hematuria.

- **Restrictions.** This is a very low-carbohydrate, low-fiber diet. All high-carbohydrate and high-fiber foods are forbidden. The diet for the first six weeks should be highly restricted in order to bring the digestive system under control, reduce inflammation, control constipation or diarrhea, reduce yeast overgrowth, control blood sugar, and begin healing the intestinal mucosa. The "die-off" of yeasts can cause symptoms to worsen over the initial short term. Supplement with the probiotics in all cases to restore the beneficial intestinal bacteria. Supplement with enzymes. Foods from the acceptable list may be added one at a time to test for sensitivity only after symptoms have been reduced. **Test a new food for at least two weeks before adding a second.** All of the food must be

Chapter 5

fresh and unprepared before cooking. **Foods not listed in this highly-restrictive Starting Diet are not acceptable—period.**

- **Meat and Animal Fats.** Eating meat and animal fats at every meal is the primary factor involved in healing intestinal diseases. This cannot be overemphasized. The complete complement of amino acids and essential fatty acids are the primary nutritional elements from which the body builds tissue and the immune system. These can only be found in meat and animal fats. Meat and animal fats also contain vast amounts of vitamins, minerals, and other nutritional elements. This diet cannot be adapted to the vegetarian lifestyle. There are no combinations of foods from vegetable sources that can provide the complete nutritional requirements necessary for healing. Most vegetable sources contain offending carbohydrates, fiber, and allergic factors that are the causes of intestinal diseases—not the cure.

- **Fish.** Eat salmon at least once a week. Cook by boiling, poaching, baking, or frying in refined coconut oil. Some people report problems with salmon which can contain rancid omega-3 fatty acids. Avoid salmon with a strong fishy smell. Canned salmon with the bones is very good as long as it is free from additives. Eat the bones in canned salmon as well. The added salt cannot be avoided. Do not eat salmon packed in soybean oil.

- **Fowl.** Eat roasted chicken thighs along with the skin. Avoid lean breast meat. Fowl injected with a salt/sugar solution is absolutely forbidden.

- **Red Meat.** Eat beef, lamb, or pork grilled, roasted, or fried. Select fatty cuts and eat all the fat. Cooking meat to medium rare is best because the enzymes are preserved. Ground meat and pork should be thoroughly cooked. Fry calf or chicken liver in refined coconut oil and include a small piece with other foods each day. Avoid overcooking liver. Bacon can be tested after healing has progressed, but the bacon must be totally free of sugars. The ingredient list on the bacon package must not contain sugar of any type. Bacon should be avoided by those who are sensitive to salt. Avoid all other prepared meats, injected chickens and turkeys, and all seasonings. Fried pork skins are an acceptable snack food and can be used with eggs to "bread" other meats and vegetables. Avoid labels listing *natural flavors.* These additives could contain sugar and MSG—a nerve toxin that makes cancers incurable.

- **Eggs.** Eggs are the perfect food for mankind. Eat eggs cooked any way desired. Frying in refined coconut oil or butter is best. Soft boiled is very good because the boiling kills any salmonella bacteria that may be present on the outside and also prevents oxidization of the healthy cholesterol in the yolk. Do not limit the quantity.

- **Vegetables.** Eat well-done boiled or steamed asparagus; eggplant, green or yellow string beans, red tomatoes, spinach, celery, peppers, or green or yellow zucchini squash only. Red, gold, and yellow peppers may be better than the standard green peppers. These must be well cooked. They are good stir-fried in refined coconut oil. Make a nice lunch or dinner by adding cubed precooked meat or leftover meat to stir-fried vegetables. Some cabbage is OK if well cooked. Try pressure cooking cabbage with green beans. Avoid the cabbage if it produces gas. The digestive system does not tolerate raw vegetables because of the fiber, but cooking reduces the fiber content. Do not drink raw vegetable juices. Canned 100% tomato juice is acceptable in moderation. Vegetables are of a minor importance, and those listed here should also be viewed with suspicion. Avoid any vegetable that gives an adverse reaction, such as cramps, pains, or intestinal gas. **Avoid all seasonings. Avoid large servings of vegetables. This diet is primarily meat, fish, and fowl with the natural fats.**

- **Cheese.** Hard cheeses are acceptable for those without a candida yeast infection. Eat only hard cheeses that list no more than 1g of carbohydrates on the nutritional label. Low-salt Swiss cheese is a very good choice that can be eaten without limit. Cheese with sodium content above 200mg per serving should be avoided, particularly by those with hypertension (high blood pressure). Steven Jenkins' *Cheese Primer* is the best reference book for a good education about cheeses. Avoid low-fat, low-cholesterol cheese, but part-skim milk cheese is acceptable. Do not eat cream cheese, cottage cheese, soft cheese spreads, dry curd cottage cheese (DCCC), or Farmer's cheese. Some good cheeses are made from sheep's milk and goats' milk, but goats' milk cheese offers nothing special or better than cows' milk cheese. Blue-veined mold cheeses may promote candida more than other types and should be limited. Do not drink or cook with kefir milk, a fermented liquid product made with active yeasts and bacteria from kefir grain and goats' or sheep's milk. Do not eat kefir yogurt.

- **Appetizers.** These must be limited to the food restrictions listed here. Acceptable examples are baked chicken wings, boiled or baked shrimp, and raw or cooked oysters. No seasonings are allowed. Do not eat chicken wings deep fried in vegetable oils.

Chapter 5

- **Desserts and Snacks.** This diet program contains no desserts in the common sense of the word. The limited acceptable foods for snacking are fried pork skins, hard cheeses, or one of the appetizer foods. Common high-sugar desserts are some of the primary causes for IBD. Most IBD sufferers will blindly follow any diet that allows desserts, such as yogurt, fruit, and honey that are forbidden on this diet program. Yogurt, fruit, and honey prevent healing and cause repeated flares. Most people continue to suffer because they refuse to admit they are carbohydrate addicts. Special homemade dips can be avocado, salsa, or mayonnaise. Commercial products are forbidden.

- **Recipes.** Do not cook using recipes. Throw your cookbooks away. Do not make soups or use thickeners of any kind. Cook chicken by baking without any additives, frying in butter or coconut oil, or grilling without any additives. Boiled chicken is OK, but it does not taste as good. Butter and coconut oil must not have any sugar added. Do not use spices. Cook fish the same way as chicken. You may choose to boil salmon, which is very good. Cook red meat the same way as chicken. A slow cooker or crock pot is great for all meats. Beef 7-bone roast is wonderful in the crock pot. Beef, lamb, or pork shoulder roasts are great. Add some diced tomatoes and let it cook for at least 5 hours. Serve with a small bowl of the fat drippings as a dip for the meat. Never eat lean meat without added fat. The pressure cooker is best for vegetables because the higher temperature breaks down the fiber. Boiling is OK if continued until the vegetables are soft or well done. During a flare, eat only green beans until your digestion improves; then add other vegetables one at a time to test. Do not eat a vegetable-only meal. Eat mostly meat, fish, or chicken with lots of fat. Eat small portions of vegetables.

- **Artificial Sweeteners.** The best sweetener is Necta Sweet, which contains sodium bicarbonate and sodium saccharin. It comes in a small bottle with a red and white label and a red cap. Clear liquid Stevia Glycerite without FOS is acceptable. Splenda® (sucralose) is acceptable but not encouraged. Use all artificial sweeteners sparingly.

- **MCT Oil.** Medium Chain Triglycerides (MCTs) are special oils for those who are unable to or have difficulty with digesting or absorbing conventional fats. MCT consists of very stable C8 caprylic and C10 capric saturated fatty acids extracted from coconut oil. MCTs are more readily hydrolyzed and absorbed than conventional food fat. MCTs require less enzymes and bile acids for digestion. MCT oil is intended for those with deficient intraluminal hydrolysis of fat (decreased pancreatic lipase, decreased bile salts), deficient mucosal fat absorption (decreased mucosal permeability, decreased absorptive surface), or deficient lymphatic transport of fat (i.e., intestinal lymphatic obstruction). These deficiencies are common in people with IBD. MCT oil can be used for low-temperature cooking or adding to foods in place of other oils. MCT oil gives quick energy, does not increase body fat, and does not increase serum cholesterol levels. MCT oil metabolizes without bile and thereby allows the gallbladder to rest.

- **Coconut and Coconut Oil.** Eat small amounts of unsweetened flaked or shredded coconut as a snack, if desired. Cook with refined coconut oil. Super Wal-Mart® stores have LouAna®, a refined coconut oil that is very good. The flavor of unrefined coconut oil is much too strong. Do not eat processed coconut flour because it is the concentrated bad portion of coconut consisting of fiber and carbohydrates. Coconut oil is 62% medium chain triglycerides (MCTs).

- **Fats and Dips.** All natural animal fats are highly recommended and should be eaten. Butter is very good, and extra-light olive oil is acceptable. Pure butter fat is called *ghee*. It is the most acceptable butter fat, but the extra cost is not justified. Saturated fat is healthy and necessary to build healthy body cells. Avoid unhealthy polyunsaturated vegetable, seed, and grain oils made from corn, soybean, canola, safflower, sunflower, cottonseed, almond, apricot, grapeseed, peanut, poppyseed, rice bran, sesame, teaseed, tomato seed, walnut, and wheat germ because they are very high in inflammatory omega-6 fatty acids. Spectrum® *Canola Mayonnaise* is acceptable, or make homemade mayonnaise as a dip for meat, fish, chicken, or steamed vegetables. Place one tablespoon of lemon juice and one raw egg in a blender. Wash the egg and hands well before cracking to avoid salmonella bacteria contamination. (Salmonella bacteria are found on the outside of the egg's shell, not inside of the egg.) Start the blender on high speed. Add one cup of extra-light olive oil very slowly in a fine steam. Take at least 60 seconds or longer to add the oil. Refrigerate and limit the shelf life. The special mayonnaise should be eaten in moderation. Start with one heaping teaspoon per meal to test.

- **Omega-3 Fatty Acids.** Take one tablespoon of Carlson's® Lemon Flavored Cod Liver Oil twice a day with a meal. Take one tablespoon 1/2 hour before a meal if you burp an unpleasant fish taste. If necessary, remove the taste from the mouth by chasing with a small amount of 100% pure tomato juice not made from concentrate. Cod liver oil has high levels of vitamins A and D. **The**

Chapter 5

importance of taking cod liver oil (EPA and DHA fatty acids) and borage oil (GLA fatty acid described below) cannot be overemphasized. **This combination is absolutely required.**

- **Omega-6 GLA Fatty Acid.** Supplement with omega-6 gamma-linolenic acid (GLA) by taking one capsule of borage oil daily. The body does not produce essential fatty acids—they can only be found in food. GLA and omega-3 fatty acids together produce prostaglandins, called *E1 series*. This type of prostaglandin helps reduce inflammation and aids in digestion. Do not take any omega-6 oils such as flaxseed or primrose oil. **Avoid all other vegetable omega-6 oils from grains, nuts, and seeds.** Solid scientific research shows omega-6 fatty acids are highly inflammatory and should never be eaten by anyone with a bowel disease, heart disease, arthritis, or any other autoimmune disease. Everyone should seriously limit these omega-6 fatty acids. Dr. Robert C. Atkins' *Age-Defying Diet Revolution* and Dr. Michael Eades' *Protein Power Lifeplan* both have long sections describing the unhealthy effects of these oils.

- **Calcium.** The **Starting Diet** excludes cheese, the major source of calcium, so a calcium/magnesium tablet can be taken in this phase. Avoid taking excessive amounts of calcium from supplements later in the diet. Calcium has a serious negative ramification in that it depletes magnesium from the body. Seventy-five percent of the people in the United States eat a diet that is deficient in magnesium. This low-carbohydrate diet allows a generous quantity of hard cheeses that are very high in calcium. The daily multivitamin tablet also has calcium; therefore, do not supplement with separate calcium tablets if generous amounts of cheese are consumed. Doctors commonly prescribe "calcium channel blocker" drugs to treat heart disease because the excess calcium to magnesium ratio in the cells of the heart muscle causes heart disease.

- **L-Tryptophan and P5P. <== Very Important.** Emotions and mood swings, such as excitement, anxiety, and depression can have very negative effects on the body, including high blood pressure, high pulse rate, poor digestion, constipation, or diarrhea. The combination of these supplements is a **major breakthrough in digestive health.** It works by calming the digestive system, hormones, and all autoimmune diseases. The combination of L-Tryptophan (an essential amino acid) and vitamin P5P (pyridoxal-5-phosphate, the active form of vitamin B6) promotes healing of the digestive system. P5P is needed to convert L-tryptophan to serotonin which calms digestion and the immune system.

Achieving Remission In Autoimmune Diseases

Serotonin promotes sound sleep and can be the cure for insomnia, an awesome side benefit. Serotonin is a neurotransmitter involved in the regulation of mood. Adverse symptoms can include sleepiness and a tired, sluggish feeling. Take one 500 mg L-tryptophan capsule and one 50 mg P5P capsule twice a day with breakfast and dinner. Do not substitute 5-hydroxy-L-tryptophan (5-HTP) for the pure pharmaceutical L-tryptophan recommended here. The body does not react to them in the same way. Do not substitute vitamin B6 for the P5P. It is not the same. These two supplements may not be available at your local vitamin store. **Do not skip this important new alternative treatment for all autoimmune diseases.**

- **Lexapro®.** The prescription drug Lexapro® is a selective-serotonin reuptake inhibitor (SSRI) that increases the serum level of serotonin by prohibiting its reuptake by the body. The generic form is Escitalopram Oxalate. Another SSRI form is Citalopram Hydrobromide. SSRIs can provide dramatic improvements in the treatment of all autoimmune diseases. It is also an antidepressant and treats excessive worry, anxiety, and insomnia. Start at 5 mg once daily as one-half of the normal dosage and remain at this level if possible. Learn the possible adverse reactions and observe them carefully. Lexapro® and Citalopram can be addictive. Do not quit taking the drug suddenly. The proper protocol for stopping the drug is to taper off over many weeks. Take Lexapro® only with your doctor's permission. **Do not take the L-Tryptophan and P5P in combination with this drug.**

- **Magnesium.** Magnesium is important to prevent osteoporosis, high blood pressure, heart disease, heart palpitations, diabetes, asthma, chronic bronchitis, migraine headaches, muscle cramps, premenstrual syndrome, chronic fatigue syndrome, fibromyalgia, depression, and other psychiatric disorders. People are commonly magnesium deficient, so this supplement is highly recommended. Take one Magnesium Glycinate 400, 200 mg three times a day with a meal. Break the magnesium glycinate in half. Take one half about one-quarter of the way through the meal and the other half at the three-quarter point to help distribute the magnesium. KAL Magnesium Glycinate 400 is the highly-absorbed form of magnesium chelated with amino acids that can be taken at twice the dosage as other brands without causing intestinal distress. Do not substitute another brand. Magnesium is essential for a healthy heart and cardiovascular system. Reduce the amount if the bowels tend to be loose.

Chapter 5

- **Potassium.** Season foods lightly with Morton's Salt Substitute (potassium chloride) instead of regular sodium chloride table salt. Other brands are not acceptable. Strictly avoid monopotassium glutamate because it has been shown to excite and activate existing viral infections such as herpes.

- **Pantethine Coenzyme A and Pantothenic Acid Combination.** The combination of Pantethine Coenzyme A and Pantothenic Acid B-5 raises the level of coenzyme A enough to relieve discomfort from inflammation. Take two capsules each per day during periods of discomfort and one thereafter.

- **Probiotics.** A healthy balance of beneficial bacteria called *probiotics* must be restored in the intestines during the low-carbohydrate, anti-yeast diet. The most common approach is to supplement the diet with a probiotic containing live lactobacillus acidophilus, lactobacillus bulgaricus, and bifidobacterium bifidum as the better choices. Lactobacillus bulgaricus is intended to help the GI tract become more suitable for the survival and growth of lactobacillus acidophilus and bifidobacterium bifidum. Acidophilus is the primary bacterium of the small intestine, while bifidum is the major bacterium of the large intestine. Other beneficial bacteria include bifidobacterium lactis, bifidobacterium longum, and others. New evidence points to lactobacillus sporogenes as being particularly effective against intestinal and vaginal infections and should also be taken. Most should be refrigerated. Capsules and liquids are available, but care should be taken in the selection because many brands contain dead spores. Quality products are usually kept in a cooler at the health food or vitamin store. Expect some or all products to contain rice flour, potato starch, maltodextrin, and other fillers that are needed to keep the bacteria alive. Take only one tablet or one teaspoon with each meal or with water. Do not take high quantities as may be recommended on some bottles. **Sporogenes is highly recommended for everyone, even healthy individuals. It does not require refrigeration. Multidophilus® by Solaray® should also be taken. Multidophilus must be refrigerated and purchased only from a refrigerated display at a local health food store.** Buy locally and refrigerate.

- **Enzymes.** Pancreatic enzyme deficiencies can cause bowel disease because various carbohydrates and proteins are not digested. Take one multiple enzyme supplement that includes amylase, maltase, sucrase, lactase, lipase (for lipids), protease (for proteins), and any others available with each meal. Discontinue taking enzymes if they cause constipation. There is no enzyme available for those who are fructose intolerant; therefore, the only solution is to avoid all fructose. Products with carbohydrate fillers should also be avoided.

- **Bile Acid Factors.** Those who have had their gallbladders removed may have difficulty digesting fats. Bile acid factors (bile salts) can be taken as a supplement with each meal to provide the bile acids necessary for the proper digestion of dietary fats. In the small intestine, bile acids emulsify fats to aid in their absorption. Bile acid deficiency causes fat malabsorption and fatty stools (steatorrhea) as evidenced by diarrhea and floating stools.

- **Protein and L-Glutamine Drink.** Prepare a drink made with whey amino acid protein powder that is enriched with extra glutamine amino acid. The protein powder consists of a full complement of amino acid isolates that heal the body and require no digestion. Prepare the drink by blending 8 to 16 oz of reverse osmosis (R.O. with UV lamp) water or unsweetened low-sodium tomato juice with 1 heaping teaspoon (12 gm) of whey protein powder plus 1 rounded teaspoon (8 gm) of glutamine amino acid powder. Stirring vigorously with the teaspoon is sufficient. The whey protein must be specified on the carton as *isolates from cross flow microfiltration and ion-exchange, ultra filtered concentrate, low molecular weight, and partially hydrolyzed whey protein peptides rich in branched chain amino acids and glutamine peptides*. The low-carbohydrate type at 1 gm per scoop or less is best, but it should not contain more than 4-5 gm of carbohydrates per scoop. Do not substitute protein from soy, egg, casein, or any other source. Sugar or any other sweetener is unacceptable. Use the "natural flavor" without additives. This amino acid drink can be enjoyed anytime with or without a meal. Amino acids are foods that build and maintain the body. **Refrigerate whey protein powder and discard if it is old.** Whey protein powder can cause some gas and an unusually "full" feeling. Discontinue the whey protein powder if the reactions are unpleasant. Continue to take the glutamine powder. This combination of amino acids has the following healing properties:

    o Provides pain killing effects by healing the nervous system.
    o Allows absorption of body building amino acids without requiring digestion.
    o Stimulates insulin-like growth factor 1 (IGF-1) which functions similarly to insulin and enhances protein synthesis and healing.
    o Fights infections by stimulating the immune system. All immune cells are made from polypeptides of amino acids.

Chapter 5

- o Provides bone growth of protein collagen and strengthens bones. Poor digestion has been shown to cause osteoporosis and degenerative bone disease.
- o Provides all of the amino acids required to heal and grow ligaments, tendons, joints, muscles, intestinal tract, heart muscle, and all other organs of the body.
- o Prevents hypoglycemia (low blood sugar) symptoms in people with hypoglycemia or diabetes.

- **Glutamine.** Glutamine is the most important supplement for IBD. It heals the intestinal tract, especially the small intestine. Colon problems are caused by undigested food passing from the small intestine into the large intestine (colon). Take 1 rounded teaspoon (8 gm) of glutamine several times throughout the day mixed in any liquid, up to 24 gm daily if desired. Glutamine is an amazing amino acid in that it can be used by the brain as a substitute fuel in place of glucose and ketone bodies. Therefore, glutamine is very important for those who may be hypoglycemic or diabetic when on a low-carbohydrate diet such as this. Diabetics and those with other health issues should consult a doctor who is supportive of the low-carbohydrate diet approach to health. **Refrigerate glutamine powder and discard if it is old.**

- **Reverse Osmosis Water.** Drink generous amounts of room temperature or warmer reverse osmosis (R.O. with UV lamp) water throughout the day. Do not drink ice water. Do not drink "natural spring mineral" water because the mineral content can be very unbalanced. The mineral program in this diet will restore all necessary minerals to the body in the correct ratios. In some people, peppermint tea and peppermint aromatic oils (carminatives) can relax smooth muscle and relieve pain caused by cramps. Enteric-coated peppermint oil capsules are also recommended. Hot chamomile tea without any sweetener after a meal is very soothing to the stomach.

- **Candida Restriction, Drugs, and Herbs.** This **Starting Diet** is already an anti-candida diet. Do not violate the food restrictions for any reason whether you have candida or not. Those with a Candidiasis infection should always avoid fermented foods, such as soy sauce, cheese, beer, alcohol, vinegar, sauerkraut, and pickles, as well as mushrooms and sugar alcohols. These would normally be considered acceptable low-carbohydrate foods. Helpful anti-candida, anti-fungal prescription drugs are Nystatin®, Nizoral® (Azole), Sporanox® (Itraconazole), and Diflucan® (Fluconazole), but some strains of candida may be resistant. Helpful herbal products should be taken before resorting to these prescription drugs. Take

Achieving Remission In Autoimmune Diseases

olive leaf extract, one 500 mg Vcap per day, increasing to three per day or as directed. As an alternative, take oil of oregano enteric-coated time-release softgel capsules, one per day and increase to two per day with meals or as directed. (Use with caution as it may cause sudden candida die-off and diarrhea.) Or take NOW Candida Clear with Pau D'Arco, oregano oil, black walnut, and caprylic acid. Other herbal products are barberry, golden seal, and Oregon grape root. The diet, probiotics, vitamins, and minerals should be started first.

- **Nasal and Sinus Treatment.** Chronic congestion of sinuses and nasal passages can be a bacterial infection that should be treated with antibiotics, but this is generally a misdiagnosis. The symptoms may be from a yeast or fungal infection that causes itching and excess mucus. The fungi are attracted to the moist environment. Treatment is easy because fungi are quickly killed by sodium bicarbonate (common baking soda). A home remedy is more effective than commercial products. Simply obtain a 1 oz. plastic nasal rinse or squirt bottle. Buy a fresh box of baking soda. Mix 1/4 teaspoon of sodium bicarbonate with reverse osmosis water in the bottle. The solution should be boiled and cooled first to prevent bacteria growth. Spray the mixture up into the nasal passage. Sniffing in the mixture is perfectly acceptable for the bronchial tubes and lungs. The treatment should be done twice a day for 5 days. The sodium bicarbonate may sting slightly at the back of the nasal passage, but this should quickly subside. This is a mild treatment but it is very effective.

- **Vitamins.** Start a vitamin, mineral, and supplement program cautiously and minimize all products that contain any form of carbohydrate, such as sugar, yeast, rice flour, etc. The following is a list of recommended vitamin and mineral supplements and the function for each. This program will not cause any vitamin or mineral toxicity. Vitamins, minerals, and supplements should be taken for optimum health. Just meeting the minimum daily requirements is missing optimum health by a mile. The multivitamin tablet listed on the next page contains 150% of the daily value (DV) of vitamin C. **Minimize supplement brands that contain sugars or starches.**

Chapter 5

Vitamin and Mineral Dosages

Dr. Atkins' Vita-Nutrient Solution: Nature's Answer To Drugs is the best source for determining the correct vitamin and mineral dosages, therapeutic (disease curing) effects, excessive dosage amounts, and normal recommendations. He discusses the vital functions of vitamins and nutritional supplements and provides a list of diseases and complaints that the supplements can help cure or alleviate.

Supplements are listed in order of priority. The most important are at the top.

Vitamin, Mineral, or Supplement	Capsule or Tablet Size 1000 mcg = 1 mg	Breakfast	Lunch	Dinner
Multivitamin and Multimineral. Centrum Silver® w/Vit. C 150% DV	---		1	
Carlson's® Lemon Cod Liver Oil with EPA and DHA Refrigerate	Tablespoon		1	
Borage Oil with 300mg GLA Refrigerate	1300 mg		1	
Lactobacillus Sporogenes by Thorne Research®	100 mg		1	

Achieving Remission In Autoimmune Diseases

Supplement	Dose				
Multidophilus® - Lactic Flora by Solaray® - Buy Locally and Refrigerate	100 mg	1			
NOW® Colloidal Minerals Refrigerate	Tablespoons See Note 3	1 Cap			
KAL® Magnesium Glycinate 400	200 mg	1	1	1	
Betaine Hydrochloric Acid (HCL) with Pepsin	650 mg	1	1	1	
Lutein and Zeazanthin	20 mg/1 mg	1			
Mega Digestive Enzymes 10X Refrigerate	As Desired	1	1	1	
MSM	1000 mg				Midnight

The following three supplements are for cardiovascular health.

Meridian Naturals® True Niacin Time-Release for Cholesterol Treatment	500 mg				4 8:00 PM

Chapter 5

Supplement	Amount			
L-Taurine	850 mg	1	1	1
P-5-P, Pyridoxal-5-Phosphate	50 mg	1		1

Optional supplement that may be included above.

Supplement	Amount			
Chelated Calcium-Magnesium	400/200 mg	None		
Ubiquinol CoQH (100 mg Kaneka reduced CoQ10)	100 mg	1		
Strontium	340 mg	1		1
Boron Complex	3 mg			1
Zinc, Chelated	30 mg			Every other day
Copper, Sebacate 22 mg	3 mg			Every other day
Manganese	10 mg			1
Chromium	500 mcg			1
Vitamin B-50	50 mg		1	
Vitamin B12 Sublingual.	500 mcg		1	
Vitamin C	500 mg	None		

Achieving Remission In Autoimmune Diseases

Vitamin E, Mixed Tocopherols, Natural Vitamin E Refrigerate	200 IU	None	
Vitamins K1 and K2 Complex	9 mg/1 mg	1 Every other day	
Folic Acid	1000 mcg	1	
NutriCology® Germanium	150 mg		1
Biotin	5 mg	1	
Alpha Lipoic Acid	300 mg	1	
Glucosamine and Chondroitin	500/400 mg	None	
Choline and Inositol	250/250 mg	1	
Silica Complex from Horsetail Extract Herb	1000 mg		1
Iodine as Norwegian Kelp	225 mcg		1
Vanadium, Chelated	2 mg		1
Selenium	200 mcg		Every other day

Chapter 5

Pantothenic Acid B-5	500 mg	1
Pantethine Coenzyme A Time Release	300 mg	1
Octacosanol	5000 mcg	1
Phosphatidylserine	1000 mg (2)	None
L-Glutamine Powder -- 4,500 mg - 1000 grams Refrigerate	1/2 teaspoon Protein Drink	1
Ultimate Lo Carb Whey Powder - by Biochem® - Refrigerate	Heaping tablespoon Protein Drink	1
L-Tryptophan Amino Acid	500 mg	As desired w/P5P
GABA Sleep Aid	500 mg	As desired

Nighttime Leg Cramps

Take 1000 mg of MSM at night to prevent nocturnal leg cramps.

"Whole Food" Vitamins

Never eat "whole food" vitamin products. The final product is concentrated fiber and sugar that is made by dehydrating fruits, vegetables, and grass. Humans should not eat grass.

Achieving Remission In Autoimmune Diseases

Other Considerations

- **Keep a Log of Your Healing Progress.** Get a notebook and record exactly how you feel each day at the same time. List the food eaten each day. Generally, one dismisses improvements when they occur slowly and discontinues a treatment when progress is being made. Bowel diseases heal very slowly.

- **Low-Blood Sugar Symptoms.** A low-carbohydrate diet will reveal other health problems such as hypoglycemia or diabetes by low-blood sugar symptoms, e.g., headache, dizziness, blurred vision, difficulty concentrating, etc. Mix 1/2 teaspoon of glutamine in a glass of water or unsweetened tomato juice to relieve the systems. Snack on fatty meat to prevent the low-blood sugar symptoms. Don't go hungry. See your doctor if symptoms persist or are moderate to severe.

- **Fiber.** Do not take fiber supplements. Do not take psyllium seed husk which is very abrasive to the digestive system. Do not take fructooligosaccharides (FOS) fiber supplements because they consist of long chains of fructose units. Do not eat wheat bran or rice bran. Fiber is a bad dude. Dietary fiber may not be digestible by the healthy individual, but it certainly is digestible by pathogenic intestinal bacteria and yeasts. **Contrary to professional medical advice, fiber is the perfect time-release food for bad intestinal bugs and one of the worst things a person can eat for good health.** These bacteria and yeasts ferment the fiber and produce alcohol, acetaldehyde, lactic acid, acetic acid and a host of other toxic chemicals. Intestinal gas (flatulence) is a sure sign that fiber and/or sugars are being fermented. The vegetarian concept of turning the intestines into a fermentation tube is ridiculous. Avoid all whole grains, brown rice, fruit, and dried beans, as they are high in both fiber and complex carbohydrates—a double blow to the digestive system. The reason gastroenterologists recommend a high-fiber diet is based on the faulty logic of Dr. Dennis Burkitt, a British surgeon who worked in Africa more than half a century ago. Dr. Burkitt's theory that barley bread prevented irritable bowel disorders has failed. The Africans were simply showing the benefits of not eating fruit and refined carbohydrates such as sugar and flour. Fiber not only does not prevent or cure irritable bowel diseases, it actually makes them worse. Studies of many other primitive societies have proved that very low-fiber diets prevent intestinal diseases and cancer as Weston A. Price, DDS, proved in *Nutrition and Physical Degeneration,* and Arctic explorers Vilhjalmur Stefansson and Karsten Anderson proved during many

Chapter 5

years of living with the Eskimos.

- **Constipation.** Constipation or a sluggish fullness can result because the digestive system is not accustomed to a healthy diet. The common advice that a person should have one or several bowel movements each day is a **myth**. Skipping one or even two days without a bowel movement is normal. Constipation can be prevented by taking the probiotics listed along with extra magnesium. Magnesium is a natural and safe laxative when taken in excess. People are commonly magnesium deficient, so this magnesium supplement is highly recommended. Take one *KAL*® *Magnesium Glycinate 400* (200 mg) three times a day with a meal. Break the magnesium glycinate in half. Take one half about one-quarter of the way through the meal and the other half at the three-quarter point to help distribute the magnesium.

- **Perfect Poop.** The quality of the stool is a perfect gauge of digestive health. *Characteristics* may be a better word because we normally don't use the word *quality* when discussing feces. Perfect poop produced by proper digestion will be medium brown in color and sink in the bowl. The stool will retain shape and not disintegrate into fragments. The volume could be considerable and medium in diameter. The water in the toilet bowl will be relatively clear, not cloudy. There will be very little or no odor, certainly not creating an offensively smelly bathroom. Flatulence (intestinal gas) during the day will be infrequent with very little or no odor. A floating stool could indicate entrained gases from pathogenic bacteria or yeasts or the improper digestion of fats. However, undigested fats in the stool do not cause any problems whatsoever. Pancreatic enzymes can also cause more gas to be entrained in the stool and cause it to float. This is not a problem. Red color in the stool indicates blood is being released near the end of the colon or rectum. Black color indicates blood is being released farther up in the digestive tract. Gray, yellow, or green colors indicate severe digestion problems. Most people accept digestion as being OK when the stool is anywhere between raging diarrhea and chronic constipation. This is not true. "Pencil poop" indicates the colon is constricted from inflammation. This diet program will produce proper digestion as can be seen by the quality of the stool. Constipation and diarrhea will be relieved.

- **Weight Loss.** This diet will help overweight people to lose weight. If weight loss is desired, the serving size should be a piece of meat the size of the palm of your hand for each meal. Fatty cuts are OK but avoid excessive fats to reduce the calories. Chicken thighs are great but don't eat the skin. Don't buy the dry, lean cuts like breast

Achieving Remission In Autoimmune Diseases

meat. Saturated fats are very healthy and heal the body, but you must place a restriction on your caloric intake as well as your carbohydrates. Eat fats in moderation. Use very little butter on vegetates or none at all. Avoid all dressings that contain fat. Don't eat fatty snacks like cheese between meals. Snack on low-carb vegetables only.

- **Weight Gain.** If weight gain is desired, simply increase the fat by using butter generously on cooked vegetables from the acceptable list. Buy extra fatty meats and eat all the fat. Eat the skin on chicken and salmon. Eat a lot of hard cheeses, and increase the amount of protein. Underweight people should never go hungry but should eat additional amounts of the acceptable foods to maintain weight or gain as desired. Underweight people easily gain weight on this diet.

- **Prescription Drugs.** Also ask your doctor about the prescription drugs Asacol® (Mesalamine), Pentasa®, Olsalazine, and Colazal®. They are helpful (not a cure) and well tolerated with very few side effects. However, they can cause colon and rectal cramps, pain, and hemorrhoids. Asacol® is a time-release tablet that releases mesalamine in the colon. Pentasa® is a mesalamine formulation encapsulated in ethylcellulose microgranules that provides timed release of the drug more proximally in the small bowel. Clinical trials have shown that Olsalazine, a mesalamine dimer, is effective in treating mild-to-moderate colitis and maintaining its remission. Olsalazine depends on an azo-bond to prevent proximal absorption of the mesalamine and to keep it in the intestinal lumen until the azo-bond is hydrolyzed and active mesalamine is released by the enzymatic action of bacterial floras in the lower ileum and colon. Bacterial cleavage of the compound releases twice the quantity of mesalamine without any sulfonamide. The new mesalamine analogs have been shown to be effective in treatment of mild or moderately active disease and in maintenance of remission. Olsalazine is given in divided doses of 1.5 to 3.0 g/day; Asacol®, 2.4 to 4.8 g/day; and Pentasa®, 1.5 to 4.0 g/day.

- **Colonoscopy.** HalfLytely® and Bisacodyl® Tablets Bowel Prep Kit is a prescription medication that your doctor may want you to take the day before your colonoscopy. This is the most suitable choice. A colonoscopy may be beneficial for the detection of cancer but offers few other benefits. A medical diagnosis of the type of colon disease is of little value because this diet program applies to all of them. The colonoscopy prep has been known to flush pathogenic bacteria from the colon, and for this reason digestion may improve after the colonoscopy. Contracting a deadly disease from

Chapter 5

improperly cleaned colonoscopy equipment makes it a high-risk procedure. **(CNN - March 24, 2009)** -- "Thousands of veterans in South Florida may have been exposed to hepatitis and HIV because of contaminated equipment after getting colonoscopies at the Miami Veterans Affairs Healthcare System, officials announced Monday."

- **Exercise.** Light or moderate exercise is recommended, but intense exercise should be avoided because it excites the immune system. People with an autoimmune diseases, such as asthma, rheumatoid arthritis, multiple sclerosis, lupus, Sjogren's dry eye syndrome, ulcerative colitis, Crohn's disease, inflammatory bowel disease, irritable bowel syndrome, and several dozen other diseases should not jog, hike aggressively, run marathons, or participate in competitive endurance races. These kinds of activities will most certainly cause the participant to have a major flare of the autoimmune disease.

- **Energy Level.** This diet can result in a lower energy level. Weakness can be common because your body is not accustomed to burning fat for energy. Leg weakness and a little pain are typical when you begin to hike stairs or a mountain as the muscles are depleted of glycogen. This can be relieved by eating one tablespoon of refined coconut oil one-half hour before exercise. This is good for weight gain but is negative for those desiring to lose weight. Just keep going and ignore the weakness the best you can. Fat burning will kick in suddenly, and up the mountain you'll go. Weakness is typical for people who have been on a low-fat, high-carbohydrate diet in which the body is not accustomed to burning fat for energy. Marathon runners experience this switchover which they call *hitting the wall.* Exercising after breakfast is the best way to kick-start the metabolism for the day. The low-carbohydrate diet lowers the body's metabolism, and a low metabolism is excellent for longevity. A high metabolism simply wears out cells faster, causes disease, and shortens life spans. Take at least 100 mg of CoEnzyme Q10 per day to increase energy production in the cells. Magnesium is also required as described.

- **Carbohydrate Addiction.** Carbohydrates are highly addictive, and many people have lived their entire lives with an insulin rush from excessive glucose intake. Insulin is the body's most powerful hormone and makes resisting carbohydrates extremely difficult. Many would rather be sick than give up their fruit with the high levels of fructose—the second most addictive carbohydrate. Other diets falsely claim nuts, fruit, honey, and yogurt as acceptable

Achieving Remission In Autoimmune Diseases

foods for the IBD suffer. Since this advice is instantly accepted by carb addicts, they continue to struggle with repeated recurrences of their disease symptoms. Carbohydrates cause IBD; they don't cure it. There are no healthy carbohydrates for the IBD suffer. You must break these addictions to be healthy.

- **Candida Restrictions.** Avoid yeast, vinegar, mushrooms, cheese, and fermented products, including Miso, for an anti-candida diet. These may be included in the diet for those IBD sufferers without a candida or other yeast infection. However, these should still be excluded for those on the IBD **Starting Diet** because candida yeast infections can be difficult to detect—though they may be present.

Kent Going on a Brisk 20-Minute Daily Bike Ride

Chapter 5

Foods We Should Eat

- **Red Meat.** Eat red meat and natural fats, including saturated animal fats. A new study shows fresh red meat has no connection to colon cancer, but manufactured meat products increase colon cancer. A high-protein diet boosts healthy antioxidant levels, but low-protein diets induce oxidative stress. Avoid labels listing "natural flavors." These additives could contain sugar and MSG, a nerve toxin that makes cancers incurable.

- **Meat, Fish & Fowl.** Eat beef, lamb, pork, fish, seafood, fowl, and wild game of any kind. Do not cut off any of the fat—eat it all. Do not skin chicken, duck, or other fowl; eat all of the skin. Eat cold water fish (such as salmon) which are high in omega-3 essential fatty acids. Eat animal protein and animal fats at every meal.

- **Fried Pork Rinds.** Fried pork skins are deep fried in their own fat and sold as fried pork rinds. They would be considered an awesome snack food except for the fact that the raw skins are heavily salted to prevent the growth of bacteria during shipping, handling, and storage prior to cooking. The unhealthy salt is heavily concentrated in the cooked rinds. Some of the salt can be removed by dunking or rinsing in hot water and eating quickly before they get soggy, but a lot of unhealthy salt remains. Eat them in moderation and completely avoid them if you have high blood pressure.

- **Saturated Fat.** Eat fat, including saturated fat. The North American Indians ate pemmican, a mixture of dried, crushed, and shredded meat mixed 50/50 with the animal fat to yield a food product that provided 70% of its calories from fat. Dried berries were sometimes added. This mixture would keep for many years. Eskimos lived all winter on nothing but caribou meat. They prepared a mixture using 80% fat and 20% caribou meat. Explorers Vilhjalmur Stefansson and Karsten Anderson found their health to be excellent. *Strong Medicine,* by Dr. Blake F. Donaldson, MD., is a book about the Inuit-style meat-only Eskimo diet.

- **Eggs.** Eggs are highly recommended as the perfect food.

- **Coconut Oil.** Fry in coconut oil or butter and use both generously in recipes.

- **Vegetables.** Eat well-done boiled or steamed asparagus; eggplant, green or yellow string beans, red tomatoes, spinach, celery,

Achieving Remission In Autoimmune Diseases

peppers, or green or yellow zucchini squash only. Red, gold, and yellow peppers may be better than the standard green peppers. These must be well cooked. They are good in a stir fry using refined coconut oil. Make a nice lunch or dinner by adding cubed precooked meat or leftover meat to stir-fried vegetables. Some cabbage is OK if it is well cooked. Try pressure cooking cabbage with green beans. Avoid the cabbage if it produces gas. The digestive system does not tolerate raw vegetables because of the fiber. Cooking reduces the fiber content. Do not drink raw vegetable juices. Canned 100% tomato juice is acceptable in moderation. Vegetables are of minor importance, and those listed here should also be viewed with suspicion. Avoid any vegetable that gives an adverse reaction, such as cramps, pains, or intestinal gas. **Avoid all seasonings. Avoid large servings of vegetables. This diet is primarily meat, fish, and fowl with the natural fats.**

- **Avocados.** Eat avocados fresh or as a dip for vegetables. Avoid prepared dip with additives. Limit quantity because they are high in fiber.

- **Cheese.** Hard cheeses are acceptable for those without a candida yeast infection. Eat only hard cheeses which list no more than 1g of carbohydrates on the nutritional label. Low-salt Swiss cheese is a very good choice and can be eaten without limit. Cheeses with the sodium content above 200mg per serving should be avoided, especially by those with hypertension (high blood pressure). Steven Jenkins' *Cheese Primer* is the best reference book for a good education about cheeses. Avoid low-fat, low-cholesterol cheese, but part-skim milk cheese is acceptable. Do not eat cream cheese, cottage cheese, soft cheese spreads, dry curd cottage cheese (DCCC), or Farmer's cheese. Some good cheeses are made from sheep's milk and goats' milk, but goats' milk cheese offers nothing special or better than cows' milk cheese. Blue-veined mold cheeses may promote candida more than other types and should be limited. Do not drink or cook with kefir milk, a fermented liquid product made with active yeasts and bacteria from kefir grain and goats' or sheep's milk. Do not eat kefir yogurt.

- **Omega-3 Fatty Acids.** Supplement with omega-3 essential fatty acids by taking Carlson's® Lemon Flavored Cod Liver Oil. Start with one tablespoon twice a day. Avoid flaxseed oil because it contains more inflammatory omega-6 fatty acids than essential omega-3 fatty acids. Flax is a poor source of omega-3 fatty acids. Your body must convert the shorter ALA fatty acid found in flaxseed oil into EPA and DHA fatty acids before you will receive major benefits,

Chapter 5

something most of us don't do well.

- **Borage Oil.** Supplement with omega-6 gamma-linolenic acid (GLA) by taking one borage oil capsule per day. The body cannot produce essential fatty acids—you must get them from the food you eat. Your body uses GLA and omega-3 fatty acids to make E1 series prostaglandins which help reduce inflammation, aid digestion, and help regulate your metabolism. Avoid all other omega-6 vegetable oils as found in nuts, seeds, and grains.

- **Vitamins and Minerals.** Supplement with a complete vitamin, mineral, enzyme, probiotic, and amino acid program as described. Avoid extra vitamin D supplementation. The cod liver oil has about 500 IU of vitamin D per teaspoon. Don't be concerned about vitamin D toxicity as erroneously claimed. You would have to take 100 times this amount for several months in order to cause toxicity. "In adults, taking 1250 µg (50,000 IU)/day for several months can produce toxicity." *The Merck Manuals.*

- **Reverse Osmosis Water.** Install a reverse osmosis water (R.O.) system with ultraviolet (UV) lamp for all drinking and cooking. The UV lamp kills all viruses and bacteria. Avoid domestic water, which contains chlorine and fluorine. Avoid mineral, natural, or spring water sold in stores as many contain undesirable minerals and contaminants. Some have been pulled from the shelves because of these contaminants.

- **Coffee, Tea & Soft Drinks.** Weak regular coffee and black teas are acceptable. Several cups of regular black coffee a day are well tolerated by most people. Don't drink decaf coffee or decaf black tea because toxic chemicals are used to remove the caffeine. Peppermint tea is preferred. A small amount of lemon or lime juice squeezed from a small wedge of fresh lemon or lime can be added to water or tea. Ginger root tea is great. Cut a piece of ginger root about the size of the small fingertip. Remove the skin. Chop and squeeze with a garlic press. Put the juice and pulp in a cup, French press, or tea ball and add boiling water for two minutes. Strain out the pulp. Ginger has a spicy flavor and is very soothing to the stomach and digestive tract. Diet sodas containing aspartame and regular sodas are absolutely forbidden. Do not drink apple cider vinegar because the health claims are unfounded. The malic and acetic acids in apple cider vinegar can burn the throat and promote bacterial vaginosis (bacterial overgrowth in the vagina).

Achieving Remission In Autoimmune Diseases

- **Beware of Bacteria-Contaminated Coffee.** Many coffee shops and restaurants use a fresh-brewed coffee dispenser that contaminates the coffee with pathogenic bacteria. These bad bacteria upset healthy digestion without the consumer knowing the source of the food poisoning. The worst type of dispenser can be recognized as an insulated tank with a pump on the top. The internal parts of the pump are rubber or plastic diaphragms and tubes that cannot be cleaned. Most coffee shops and restaurants make no attempt to clean the pump mechanism. Rubber and plastic parts easily harbor bad bacteria that continue to multiply. The coffee in the pump also cools easily. The temperature of the coffee is not high enough to sterilize the parts. The best coffee dispenser is the type with a spherical clear glass pot that is very easy to clean and inspect. The coffee filter mechanism should be a stainless steel cone which is also easy to clean. Ask your coffee shop or restaurant about its coffee brewing and dispensing systems. Inspect the clear glass serving pots for cleanliness. Don't drink coffee that is brewed and dispensed from a poorly-designed system.

- **Coconut.** Eat and cook with refined or virgin coconut oil, and eat unsweetened shredded coconut. Try both together at the same time for a taste treat. Do not eat processed coconut flour because it is the concentrated bad portion of coconut consisting of fiber and carbohydrates.

- **Nuts and Seeds.** Avoid most nuts and seeds because they contain high levels of omega-6 fatty acids. Hazelnuts (filberts), macadamia nuts, and pine nuts (pinion, pinon, pinyon or pignolia) are great snacks with the lowest amount of omega-6 fats. Limit quantities to test for a reaction. All other nuts and seeds are forbidden. All species of nuts are not healthy as many claim.

- **Candida Restrictions.** Avoid yeast, vinegar, mushrooms, cheese, and fermented products, including Miso, for an anti-candida diet. These may be included in the diet for those IBD sufferers without a candida or other yeast infection. However, these should still be excluded for those on the IBD **Starting Diet** because candida yeast infections can be difficult to detect—though they may be present.

Chapter 5

Foods That Are Absolutely Forbidden

- **Detox Diet Plans.** Never participate in any of the popular detox diet programs. All of these programs recommend many foods that are very harmful to people with food-caused autoimmune diseases. These plans typically forbid eating the healthy foods listed here. Detox diet plans are always low-fat, which means they are high in harmful carbohydrates. The *detox* concept is a fraud and a scam. The body does not build up toxins except in cases of rare trace metals such as mercury or lead. The detox diet programs will not remove poisonous metals from the body. A standard scientific laboratory should be used to test for mercury or lead poisoning. Most alternative medicine tests for toxins and food allergies are a scam as well. Detox diet plans also run the risk of causing leaky gut syndrome, the major cause of autoimmune diseases.

- **Sugar and Sweets.** Do not eat sugar in any form. Sugar raises the level of free-radicals and blood insulin which causes heart disease and diabetes. Do not eat corn syrup, fructose, honey, sucrose, maltodextrin, dextrose, molasses, rice milk, soy milk, grape juice, fruit juice, brown rice syrup, maple syrup, date sugar, cane sugar, corn sugar, beet sugar, succanat, or lactose. Do not eat candy, cookies, ice cream, cakes, dates, crackers, soft drinks, or yogurts, which are all high in carbohydrates. I have described diabetes, heart disease, and cancer as *Carbohydrate Addicts' Syndrome*.

- **Honey.** Do not eat honey. Honey is pure carbohydrate sugar consisting of fructose and glucose, sucrose, maltose, isomaltose, maltulose, turanose, kojibiose, erlose, theanderose, and panose. It is very confusing why people worship honey. Many people are wrong on this point. They claim it is acceptable because ancient cavemen may have eaten it and somehow became adapted to honey as a healthy food. That is an incorrect conclusion. Ancient Paleo men may have eaten it once in a lifetime but probably not. They didn't want to be stung by bees any more than we do, and they didn't have any protective netting.

- **Sugar Alcohols.** Many of the "sugar free" sweeteners are classified as *sugar alcohols*. Sugar alcohols affect the blood glucose levels less dramatically than regular table sugar, but they quickly add up to too many carbohydrates. They contain a little more than one half the amounts of carbohydrates as an equal amount of table sugar. Common sugar alcohols are mannitol, sorbitol, xylitol, maltitol, maltitol syrup, galactitol, erythritol, inositol, ribitol, dithioerythritol, dithiothreitol, and glycerol, as well as

Achieving Remission In Autoimmune Diseases

hydrogenated starch hydrolysates that are found naturally in fruit. These sugar alcohols generally have unpleasant side effects such as abdominal discomfort and bloating. They also have a laxative effect.

- **Processed Meats.** Do not eat processed meats, such as sausage, hot dogs, ham, deli meats, injected turkey, and injected chicken due to the added chemicals and sugar and salt content.

- **Starchy Vegetables, Lettuce, and Spinach.** Limit carbohydrates in all forms except low-starch vegetables. Avoid those starchy vegetables that grow below the ground, such as potatoes, yams, turnips, beets, radishes and carrots. Avoid pumpkin and Winter squash. Never eat lettuce and raw spinach because of the risk of pathogenic bacteria contamination. Cooked spinach is OK. Salad bars are a high risk for food poisoning.

- **Fruit and Fruit Juices.** Do not eat fruit of any kind because of the sugar, and do not drink fruit juices. Fructose (fruit sugar) has been linked to insulin resistance, the primary cause of diabetes. Fruit promotes the growth of pathogenic intestinal bacteria, candida yeast, and fungi. Do not drink apple cider vinegar because the health claims are unfounded. The malic and acetic acids in apple cider vinegar can burn the throat and promote bacterial vaginosis.

- **Margarine and Trans Fats.** Do not eat margarine, commercial mayonnaise, or any product that contains hydrogenated oils (trans fats).

- **Omega-6 Fatty Acids.** Strictly avoid omega-6 polyunsaturated vegetable, seed, or grain oils made from corn, soybean, canola, safflower, sunflower, cottonseed, almond, apricot, grapeseed, peanut, poppyseed, rice bran, sesame, teaseed, tomato seed, walnut, and wheat germ. Do not take omega-6 oils such as flaxseed or primrose oil. The omega-6 oils cancel the benefits of the good omega-3 fat. One beneficial omega-6 fatty acid is gamma-linolenic acid (GLA) which can be obtained by supplementation with borage oil. Solid scientific research shows omega-6 fatty acids are highly inflammatory and should never be eaten by anyone with bowel disease, heart disease, arthritis, or any other autoimmune disease. Healthy people should seriously limit these omega-6 fatty acids. Dr. Robert C. Atkins' *Age-Defying Diet Revolution* and Dr. Michael Eades' *Protein Power Lifeplan* both have long sections describing the unhealthy effects of these oils.

Chapter 5

- **Milk.** Do not drink cows' milk. Lactose in cows' milk is one of the most allergenic foods, and the symptoms are commonly described as *lactose intolerance*. Milk does not prevent osteoporosis. Do not drink goats' milk, rice milk, soy milk, lactaid milk, acidophilus milk, almond milk, or nut milks of any kind. Do not drink or cook with kefir milk, a fermented liquid product made with active yeasts and bacteria from grains and goats' or sheep's milk. Do not eat kefir yogurt. "Likewise, higher intakes of total dietary calcium or calcium from dairy foods were not associated with decreased risk of hip or forearm fracture." ("Milk, dietary calcium, and bone fractures in women: a 12-year prospective study". *American Journal of Public Health.* 1997 Jun;87(6):992-7.)

- **Heavy Whipping Cream.** Do not use cream or heavy whipping cream because it has an unacceptable level of lactose called *buttermilk* after separation from the fats. Commercial heavy whipping cream also has thickening additives such as carrageenan which may promote malignancy and inflammation in the gastrointestinal tract and cause cancer.

- **Pre-Whipped Cream and Toppings.** Do not eat pre-whipped cream or low-fat, non-dairy whipped cream because these contain sugar or trans fats. See the list of typical store foods containing heart-clogging trans fats.

- **Yogurt.** The acidophilus in yogurt is simply killed by stomach acid and does not provide the desirable probiotics for the intestinal tract. The lactose in yogurt feeds candida yeast and pathogenic bacteria. All yogurts are bad food. Goats' milk yogurt is also bad food.

- **Soy Products.** Do not eat any soy products. Fermented foods are not acceptable for those with yeast overgrowth infections. Do not use soy protein powders, but whey protein powders are acceptable. Soy protein is missing several of the amino acids, one of which is classified as an essential. Do not eat soy protein chips or cereals. Tofu made from soybeans has been shown to shrink the brain and cause cognitive impairment (brain fog).

- **Wheat, Corn, and Other Grains.** Do not eat any wheat products or other grains, such as corn, oats, rye, rice, barley, millet, kamut, or spelt. Corn is missing three of the essential amino acids necessary for good health. Grains are poor sources of protein. Do not eat any bread or other flour products. Grains are the most allergenic of all foods. Multiple sclerosis, lupus, and rheumatoid arthritis are rare in populations where no grain products are

Achieving Remission In Autoimmune Diseases

consumed. The Paleolithic (hunter-gatherer) diet is an example.

- **Packaged Foods and Snacks.** Do not eat breakfast cereals, pancakes, waffles, bread, biscuits, tortillas, taco shells, bagels, pasta, noodles, corn chips, pop corn, croutons, spreads, dressings, desserts, soups, soy snacks, rice snacks, candy, cakes, or pastries because they usually contain partially hydrogenated vegetable oils, omega-6 fatty acids, MSG, chemical thickeners, colorings, and preservatives; they are all very high in carbohydrates.

- **Potatoes and Yams.** Do not eat any potatoes, sweet potatoes, yams, French fries, or potato chips. French fries are not only very high in unhealthy carbohydrates, but they're generally cooked in hydrogenated soybean oil or other rancid vegetable oils that are known to generate cancer-causing chemicals at high temperatures and are high in inflammatory omega-6 fatty acids. Always fry in coconut oil or butter.

- **Bananas and Citrus.** Do not eat bananas, oranges, or grapefruits; they are very high in carbohydrates.

- **Legumes.** Do not eat beans and legumes because of their high carbohydrate levels. Limit peanuts because they are a legume and contain a very unhealthy fat.

- **Bad Cheeses.** Do not eat soft, dark orange cheeses that are made from partially-hydrogenated oils (trans fats). Avoid processed cheese foods which contain corn oil instead of butterfat. Do not eat cream cheese or cottage cheese since they contain lactose and other sugars. Contrary to popular belief, cottage cheese has been proven to cause bone loss and osteoporosis.

- **Smoking.** Stop smoking—but stop eating sugar first. Sugar will destroy your health faster than smoking.

- **Low-Fat Products.** Do not eat any product labeled *low-fat*.

- **Hydrogenated Oils.** Do not eat anything containing hydrogenated oils, called *trans fats*. Read every label.

- **Low-Cholesterol Products.** Do not eat any product labeled *low cholesterol*.

Chapter 5

- **Carbohydrates.** Avoid carbohydrates from all sources except low-starch vegetables.

- **Starch Blockers.** Do not take starch-blocker supplements or drugs; they promote fermentation of the starch by yeast in the colon. The undigested starch also provides a high-energy food source for pathogenic bacteria and toxin-producing fungi.

- **MSG.** Avoid monosodium glutamate (MSG) and the dozens of flavors and seasonings which have deceptively hidden the MSG glutamic acid ingredient.

- **Carrageenan.** Avoid carrageenan, a gum extracted from red seaweed and used as a fat substitute to thicken many food products. Unfortunately, carrageenan is used in many otherwise acceptable low-carbohydrate foods such as processed meats and heavy whipping cream. Carrageenan may promote malignancy and inflammation in the gastrointestinal tract and other cancers.

- **Fast-Food Restaurants.** Avoid fast-food hamburgers that are bulked-up with soy protein and/or cooked in trans fats or polyunsaturated fats. Burgers that are labeled as 100% beef and cooked by grilling or frying in their own fat are the best. Avoid French fries as well because the frying oils are either trans fats, polyunsaturated fats, or fat which has become rancid from overuse. Grilled chicken breast is an acceptable choice. Don't eat the bun. Don't eat anything coated or any deep fried foods.

- **Restaurant Chicken Wings.** Restaurants ruin most of the good foods they touch. Chicken wings are a good example of an awesome food that is destroyed by deep frying in trans fats or unhealthy omega-6 polyunsaturated fats such as soybean, corn, safflower, or peanut oil. These fats easily become rancid which makes them even unhealthier. Baked chicken, fish, and red meat are OK if they are not heavily salted or treated with tenderizers such as MSG. You can ask the restaurant manager but he may lie.

Household Health Issues

- **Safety of Plastics.** Avoid food containers made from plastics that have been shown to release harmful chemicals into the food or water stored in the containers. The bottoms of the containers should have a triangle mark with a number. Numbers 1, 2, 4, and 5 are least harmful. Avoid all others. Glass, ceramic, stainless steel,

Achieving Remission In Autoimmune Diseases

and cast iron are more acceptable containers for food.

- **Microwave Ovens.** Avoid the use of microwave ovens except for heating plain water. Microwave energy destroys enzymes and vitamins. It also breaks weak molecules into undesirable elements. The energy can cause the unhealthy cross linking between molecules. Microwaving starches such as potatoes and other carbohydrates like bread creates high levels of acrylamide, a chemical proved to cause cancer in laboratory animal tests.

- **Soap, Shampoo, Body Cream, Powder, and Oils.** Use facial make-up, body creams, lotions, oils, and body powder sparingly. Do not use those containing wheat protein, sugars, fruit juices, or omega-6 fatty acids as found in vegetable, nut, and seed oils. Coconut oil, cocoa butter, palm oil, olive oil, fish oils, and natural animal fats are acceptable but are still prone to rancidity. The term *natural* does not necessarily indicate that the product is acceptable. Products with natural soybean oil, safflower oil, and other seed oils are absolutely dreadful oils because they are highly inflammatory. Use make-up sparingly. The following are some acceptable products.

- **Hair Shampoo:** Shikai® Everyday Shampoo (or other brands that do not contain sodium lauel/laureth sulfates and have all natural ingredients). A second possible product is Cal Ben® Shampoo and Conditioner.

- **Hair Conditioner:** Shikai® Everyday Conditioner or any other brand that does not contain offensive ingredients. A second possible product is Cal Ben® Shampoo and Conditioner.

- **Hand and Body Soap:** Cal Ben® Pure Soap or Clearly Natural® Glycerin Soap, Unscented.

- **Hand, Foot, and Body Moisturizing Cream:** Refined coconut oil is best. Other choices are NOW® 100% Pure Shea Butter or NOW® Soft Cocoa Butter with Jojoba Oil.

- **Hand Dishwashing Liquid Soap:** Ultra Dishmate by Earth Friendly Products® or any natural product that does not contain sodium laurel/laureth sulfates. A second possible product is Cal Ben® Seafoam Dish Glow.

- **Clothes and Bedding Laundry Soap:** The best soap is made by Seventh Generation®. It contains enzymes that effectively remove

Chapter 5

protein and starch residue from clothing and bedding. It is hypo-allergenic.

- **Clothes and Bedding Fabric Softener:** The fabric softener is made by Seventh Generation®. It contains canola oil that effectively softens clothing and bedding and is hypo-allergenic.

Flare-Ups and Prescription Drugs

Catching the flu or a cold can also cause intestinal autoimmune diseases to flare. This is temporary and not brought about by the diet program. The flare passes as the cold passes, but taking prescription drugs during a flare can moderate the symptoms. Eating forbidden foods create a continuous flare condition until all of these foods are removed from the diet.

Prescription drugs mask the bad effects caused by forbidden foods; therefore, foods beyond the **Starting Diet** cannot be tested while taking the drugs. Foods on the forbidden list should never be tested at any time as they will almost certainly cause a flare. Upon approval by the prescribing physician, drugs can be cut back slowly as healing progresses.

Anti-Inflammatory Herbal Supplements

5-Loxin Boswellia Serrata Extract® 150 mg. Boswellia Serrata Extract has been used for centuries as an anti-inflammatory to relieve pain and discomfort of the neck, back, connective tissues, and joints; rheumatoid arthritis, osteoarthritis, Crohn's disease, ulcerative colitis, asthma, and other inflammatory bowel and autoimmune diseases. The active ingredient is acetyl-11-keto-beta-boswellic acid (AKBA). This product is standardized for a minimum of 30% (45mg) of AKBA. Research has shown it to be about as effective as nonsteroidal anti-inflammatory drugs (NSAIDS) in reducing pain, cramps, and discomfort. Daily divided doses up to 300 mg per day have been shown to be more effective than the standard recommended dose of 150 mg.

Turmeric Extract with Bioperin 900 mg. Turmeric (Curcuma longa) is an ancient spice native to India and Southeast Asia. Best known for its distinctive flavor and yellow color, it is used in curries and some prepared mustards. Other than a food additive, turmeric has been used for centuries in Ayurveda, Siddha, Unani, and other traditional medicines as a remedy for stomach and liver ailments.

Curcumin, the active ingredient in turmeric, contains a mixture of powerful phytonutrients known as *curcuminoids*. Curcuminoids have antioxidant properties which fight the damaging effects of free radical molecules in the body.

Curcuminoids may play a part in blocking a key biological pathway that damages cells and may lead to their unhealthy, unrestrained growth. They shut down nuclear factor kappa B (NF-kB), which is known to regulate expression of more than 300 genes that promote inflammatory responses and produce joint inflammation and cell damage. NSI® Turmeric Extract with Bioperine® is standardized to 95% curcuminoids.

Enzymatic Therapy® Saventaro® Max-Strength Cat's Claw® -- 20 mg - 90 Caps. Because of their antioxidant properties, the beneficial pentacyclic oxindole alkaloids (POAs) in cat's claw have been shown to help control inflammation, soothe irritated tissue of the gastrointestinal tract, and improve the immune system. Refined cat's claw must be taken because the undesirable tetracyclic oxindole alkaloids (TOAs) have been removed.

Chapter 5

- Saventaro® is patented, clinically studied, and guaranteed to benefit the immune system.
- It's the first and only Cat's Claw product made from the root of the Uncaria tomentosa plant in which beneficial POAs (pentacyclic oxindole alkaloids) are 200% more concentrated than in the bark. Saventaro is standardized for a minimum of 1.3% POAs.
- It's the first and only product guaranteed 100% free of TOAs (tetracyclic oxindole alkaloids). Contamination by as little as 1% TOAs causes a 30% reduction in the acquired immune benefits that POAs provide.

Quercetin (500 mg) and Bromelain (375 GDU). Quercetin is a powerful antioxidant that also supports the immune system. The natural antihistamine may help to relieve allergy symptoms, and its anti-inflammatory properties may help to relieve symptoms of asthma, arthritis, inflammatory bowel diseases, and other autoimmune diseases.

Illustration of a Diseased Colon with Polyps and a Large Cancerous Tumor

Achieving Remission In Autoimmune Diseases

Complete List of Autoimmune Diseases

Achlorhydra Autoimmune	Active Chronic Hepatitis	Addison's Adrenal Disease
Alopecia Areata	Anti-Phospholipid Syndrome	Atopic Allergy
Autoimmune Atrophic Gastritis	Celiac Disease	Chronic Fatigue Syndrome
Chronic Obstructive Pulmonary Disease (COPD)	Crohn's Disease	Cushing Hypercortisol Disease
Dermatomyositis	Diabetes Melitus Type 1	Discoid Lupus
Erythematosis	Fibromyalgia	Goodpasture Lung & Kidney Disease
Grave's Disease Hyperthyroid	Hashimoto's Hypothyroid	Hepatitis, Active Chronic
Idiopathic Adrenal Atrophy	Idiopathic Thrombocytopenia	Interstitial Cystitis Bladder
Lambert-Eaton Muscle Nerve	Lupoid Hepatitis	Lupus Erythematosus
Lymphopenia	Migraine Headache	Mixed Connective Tissue Disease

Chapter 5

Multiple Sclerosis	Pemphigoid	Pemphigus Vulgaris
Pernicious Anema	Phacogenic Uveitis	Polyarteritis Nodosa
Polyglandular Auto. Syndromes	Polymyalgia Rheumatica	Primary Biliary Cirrhosis
Primary Sclerosing Cholangitis	Psoriasis	Raynaud's Blood Vessel Disease
Reiter Arthritis Spine Disease	Relapsing Polychondritis	Rheumatoid Arthritis
Schmidt's Syndrome	Scleroderma - Crest Syndrome	Sjogren's Dry Eye Syndrome
Sympathetic Ophthalmia	Takayasu's Arteritis	Temporal Giant Cell Arteritis
Thyrotoxicosis	Type B Insulin Resistance	Ulcerative Colitis
Vitiligo	Vulvodynia Vulva	Wegener's Granulomatosis

Achieving Remission In Autoimmune Diseases

One of My Favorite Gourmet Autoimmune Meals

This meal is a good way to use leftover meat, such as beef, pork, lamb, chicken, or turkey. It is especially good for adding moisture and fat to dry lean meat such as turkey breast.

Ingredients:

Refined coconut oil (LouAna® at Super Wal-Mart® or equal)
Cooked meat, such as beef, pork, lamb, chicken, or turkey
Fresh green and/or yellow zucchini
Eggs, one or two eggs per serving
Coarsely grated or sliced Swiss cheese and cheddar cheese

Directions:

Dice the meat into 1/2" (12mm) cubes. Slice the zucchini lengthwise and cut into 1/4" (6mm) slices. Sauté the meat and zucchini in coconut oil on low heat until zucchini is soft but not mushy.

Break the eggs into a bowl and scramble until yolks are broken. Pour the eggs over the meat and zucchini and stir the mixture until the eggs are cooked.

Spread the cheese over the top of the meat, zucchini, and eggs and cover with aluminum foil until cheese is melted.

Serve and eat while still hot.

Serve with hot peppermint tea in the winter. It settles and calms the digestive tract.

Chapter 5

Welcome to the University of Higher Brainwashing.
The Big Bang created the universe.
You will believe everything I say.
You evolved from pond slime.
Do not think or reason.
Truth does not exist.
Question reality.
You are asleep.

Absolute Truth Exposed

Volume 1

Chapter 6

Brainwashing, Mind Control, and the Deception of Society

Typical Support Group Meeting of
Victims of Brainwashing

Chapter 6

Brainwashing, Control, Deception, Psychiatry, Psychology, Psychotic, Sociology, Sociopath, Schizophrenia, Anorexia, Bulimia, Depression, Obsessive-Compulsive, Paranoia, Phobia, Addiction, and Other Mental and Personality Disorders.

Brainwashing of Humans Is Easy and Very Effective

Brainwashing is not new. It goes back to the dawn of civilization. The serpent brainwashed Adam and Eve in the Garden of Eden, and they succumbed very easily. They believed a lie that conflicted with the truth they had already received. Such is the nature of humans.

People will accept ridiculous policies, practices, and teachings through brainwashing techniques that are so effective they deny they have been brainwashed. They are not aware of it themselves. Brainwashing results in a mental disorder. Humans are easily taught to believe something that can quickly be proved to be absolutely false. A brainwashed person's mind typically locks up in denial after he is given the truth. Brainwashing can extend to groups of people without a limit in size or scope. Until the fall of the Soviet Union exposed the truth, millions of people were brainwashed into believing that Communism was the best economic system. Parents of brainwashed children have engaged reprogramming experts in attempts to rescue their children. This method can work to some extent with children, but rescue becomes extremely difficult when the victim is an adult who is protected by law from outside rescuers. The adult can live his entire life without ever knowing that he was brainwashed.

History has exhibited long periods of time in which the entire population believed a massive lie. The claim that the Earth was flat is a good example. Another is the claim of Galileo Galilei (1564 - 1642), a mathematician, physicist, astronomer, and

Brainwashing of Society

philosopher, that the sun was the center of the solar system—not the Earth. For centuries everyone had believed that the Earth was the center of the solar system and the universe. The Pope placed Galileo under arrest and on trial for heresy for writing his book, *Dialogue Concerning the Two Chief World Systems*. The Pope used verses such as:

> **Psalm 104:5** for his charge, "[You] [who] laid the foundations of the earth, So [that] it should not be moved forever."

Galileo was found guilty and sentenced to imprisonment which was later commuted to house arrest. His book was banned. Galileo returned to his home in Arcetri near Florence, Italy, where he continued to summarize other books that he had been working on for 40 years. Galileo went blind before his death, but his work changed the world. He has since become known as *The Father of Modern Physics*.

Arctic explorer Vilhjalmur Stefansson was severely criticized for his claim that Eskimos lived in awesome health on an all-meat diet. Nutritionists and physicians said that was preposterous; they would surely die. I guess the nutritionists thought there were orange trees on the Arctic Ice Cap.

Stefansson and Karsten Anderson made two expeditions to the Arctic in the years between 1906 and 1912. They both proposed a test and allowed themselves to be admitted to Belleview Hospital in New York City where doctors would observe them for a full year on an all-meat diet. Both men showed improvements in health. Stefansson was permitted to leave the hospital and continue the test as an outpatient. His books, reports, and test results did not change the minds of the nutritionists or physicians who continued with the centuries-old myth that fruit

Chapter 6

and vegetables are requirements in the human diet. This is the nature of brainwashing. Absolute scientific facts cannot undo the erroneous convictions of the severely brainwashed mind. A hundred years have passed and nothing has changed. This book will only help those who remain able to think for themselves and discern truths from lies. An example of severe brainwashing was found on an Internet message board where a woman wrote,

> "I mistakenly ate meat after 30 years clean.....how do I detox?
> HELP PLEASE"

Her unscientific mental condition is not uncommon. In fact, her line of thinking is spreading like a pandemic, particularly among coeds on college campuses who can be seen drenched in sweat running in nearby neighborhoods. As I ride my bike for my quick 20-minute workout, I pass them and see grimacing expressions of running in pain on their faces. One runner came into the fast food restaurant to use the bathroom every morning as I was enjoying a cup of coffee. She was a Caucasian woman approximately 30 years of age, but her sun-darkened skin gave her the appearance of a full-time lifeguard at the beach. She had an obvious case of brainwashing and a severe adrenaline addiction as evidenced by her sweat-drenched body.

These young women who run their guts out every day should ask themselves, "Where are the older women who began running 20 years ago?" If running is healthy, they should be here also or they should be riding a bike alongside of me. These runners are suffering from mass brainwashing.

At best, knees only last for 20 years of jogging a few miles each day. These women will only be 40 when they have knee replacement surgery and are told to never run again.

Brainwashing of Society

Brainwashed Humans Cannot Reverse Their Own Deception

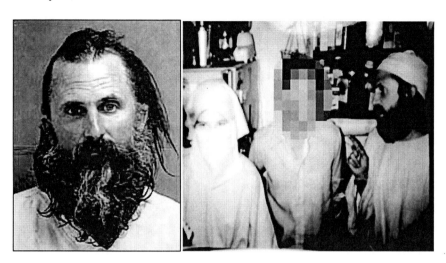

The human brain is miserably inept at correcting its depraved condition. The kidnapping of 14-year old Elizabeth Smart by Brian David Mitchell in Salt Lake City, Utah, on June 5, 2002, is a good example. She was in captivity for only nine months before she was identified and approached by police. When officers approached the teen, Sandy, Utah, Officer Bill O'Neal said, "She kind of just blurted out, 'I know who you think I am. You guys think I'm that Elizabeth Smart girl who ran away.'" Certainly anyone in her right mind would not want to stay with the freak who kidnapped and raped her, but her brainwashing was so effective that she made no attempt to escape after being held captive for only a few weeks. This is not to discredit Miss Smart in any way. Nearly every child in the same situation would have responded as she did. The case clearly shows the susceptibility of the human mind to becoming brainwashed.

Brian Mitchell brainwashed himself into believing he was some sort of deity. He had also brainwashed his wife, Wanda, to such an extent that she accepted the kidnapped girl as Mitchell's

Chapter 6

second wife. Defense attorneys attempt to make pleas for people like Brian Mitchell on the basis of insanity simply because he brainwashed himself. Certainly he was not insane simply because he thought this way.

During her captivity, Elizabeth Smart lost touch with reality and was unable to extract herself from her captor. She didn't try to escape and later gave false information to the police to protect him. Her parents carefully deprogrammed Elizabeth after her rescue because she could not free herself mentally or physically, and they were benefited by the fact that she was still a child in their custody. The same is not true for brainwashed adults because courts in the United States hold adults responsible for themselves. An adult can become hopelessly trapped by brainwashing, but the courts will not offer any outside assistance. This is not to suggest that the courts could help. Members of the court are generally brainwashed themselves.

Patricia Hearst was kidnapped in February 1974 by a neo-revolutionary group called the Symbionese Liberation Army (SLA). For the next two months—by her account—Hearst was kept in a closet and brainwashed by the small group of radicals who targeted wealthy capitalists as the ultimate enemy. Patty Hearst was a college student in Berkeley, California, and the granddaughter of publishing magnate, William Randolph Hearst. Following her brainwashing, she continued to submit to her captors and committed bank robbery for which she was convicted and sent to prison. She was later released. Her case shows that prosecutors do not understand the power of brainwashing. Patty Hearst is pictured

Brainwashing of Society

here as "Tania," in an image circulated worldwide after her kidnapping by the Symbionese Liberation Army.

A novel example of group brainwashing occurred with the Reverend Jim Jones cult. Jim Jones was born on May 13, 1931, in Lynn, Indiana, and by 1963, he had his own religious congregation, the People's Temple Full Gospel Church, located in Indianapolis. Jones led the interracial congregation—rare at the time—with faith healing, visions, and advice from extra-terrestrials. Belief in extraterrestrial aliens is a sure sign that someone is brainwashed. Jones led his brainwashed group to Brazil where he further brainwashed them into committing mass suicide. On November 18, 1978, over 900 members of his group died in an apparent mass suicide because the megalomaniac Jones convinced most of his followers to drink a cyanide mixture. Some, including Jones, were either murdered or committed suicide. Shortly before the mass suicide, US Congressman Leo Ryan was assassinated on Jones' orders. Ryan had just landed in Guyana to investigate alleged human rights abuses at Jonestown, Brazil.

The Psychological Trait That Allows Humans to Be So Easily Brainwashed

The ease with which a person becomes brainwashed is not a matter of intelligence or IQ. A scam artist can trick a person with a Ph.D. Many people with advanced degrees falsely believe that communism is a sound economic system. It's just a matter of discerning truth from error, but the human mind seems to prefer brainwashing over truth by a wide margin. People like a "sales job" which they readily accept from profit-motivated salesmen. Though this is true worldwide, it is more prevalent in English-speaking countries. Most people easily open their minds to salesmen who tell them the most far-out, nutty, illogical, stupid

Chapter 6

nonsense, but they tense-up, turn-off, and flee from sound, truthful, scientific information as presented in this book.

People are brainwashed because of their psychological arrogance. Some are so arrogant, self-centered, and proud that whatever they believe is automatically assumed to be correct. Since they make no effort to dig out the truth, they brainwash themselves. The following are just a few major examples of arrogant self-brainwashing by people with impressive credentials.

NASA Management Brainwashed Themselves and in Doing So Killed 17 Astronauts

NASA space managers allowed the Apollo 1 command module to be designed with combustible insulation on the spacecraft wiring in an oxygen-rich environment. The insulation caught fire due to a spark from a loose electrical connection, killing all three astronauts during a pre-flight test on January 27, 1967. The NASA managers were arrogant and prideful. NASA management ignored absolute truth and reality.

The new NASA space managers were brainwashed as a result of arrogance. They launched the Space Shuttle Challenger in below-freezing weather after engineers told them the rocket motor seals would leak. Scientific tests by the engineers had already proved the specified temperature limitations. The previously-fired solid-propellant rocket motors had leaked but did not explode. The temperature limit should have been raised and the seals redesigned based on the obvious reality. The rocket blew up on January 28, 1986, killing all seven astronauts. After the disaster, NASA managers tried to blame it on the engineers. NASA management ignored absolute truth and reality.

Brainwashing of Society

The replacement NASA space managers were also brainwashed as a result of arrogance. They simply dismissed the fact that insulation from the external fuel tank was coming off of the Space Shuttle Endeavour in February 2000. The same thing occurred on the Space Shuttle Columbia flight in February 2003 because the problem was ignored. The re-entry insulation on Columbia was critically damaged during launch. On re-entry from space back to Earth on January 16, 2003, the heat burned through the Shuttle. It exploded and killed all seven astronauts. The problem was still not resolved, and when the Space Shuttle Discovery was launched in July 2005, it suffered damage as well. NASA management ignored absolute truth and reality.

Two Major Nuclear Power Plant Accidents Were Caused by Brainwashing

The total destruction of the Chernobyl Nuclear Power Station in the Ukrainian Soviet Socialist Republic, USSR, on April 26, 1986, and the release of radioactive nuclear fuel at Three Mile Island Nuclear Power Station in Pennsylvania, USA, on March 28, 1979, appeared at first to have occurred for totally different reasons. This is not the case. They both happened for the same basic reason. Plant operators at both locations were arrogant, prideful, and brainwashed. They bypassed well-designed safety systems and allowed the nuclear reactions to continue instead of shutting down. They ignored reality.

Chapter 6

Other Examples of Mass Brainwashing of Society

Arrogance and brainwashing are evident when a person believes in UFOs, Bigfoot, Loch Ness Monster, alien crop circles, life on Mars, and other nonsense.

Arrogance and brainwashing are evident when a person believes saturated fat in the diet causes heart disease.

Arrogance and brainwashing are evident when a person believes birds carried around larger and larger wings for millions of generations until they finally evolved wings large enough to fly.

Arrogance and brainwashing are evident when a person believes red meat is unhealthy food.

Arrogance and brainwashing are evident when a person believes the Big Bang and Red Shift Theories.

Arrogance and brainwashing are evident when a person believes carbohydrates are healthy energy foods.

Arrogance can easily be observed in youth and students as expressed by their dress, hair, actions, and language. They do everything they can think of to be as arrogant as possible. Arrogance is one of the most rampant psychological illnesses. Arrogance is human nature. Humans are born with an abundance of arrogance. Truth and wisdom come only as a result of hard work, but young people avoid hard work. They simply succumb to the arrogance. Arrogant students suck up brainwashing like a sponge. The first step toward brainwashing a student is to tell him there is no absolute right or wrong. Students have been brainwashed into believing that absolute truth does not exist. Well, here it is in this book. His arrogant denial of the truth leaves him open to continued brainwashing.

Brainwashing of Society

Brainwashed people cannot think and reason logically. They refuse to try and refuse to consider anything that is true. Their minds are cesspools of false information.

The list goes on and on forever because error, arrogance, pride, and brainwashing are much more prevalent than wisdom. Diet and health topics are good examples of psychological brainwashing. Vegetarians are psychologically brainwashed to such an extent that switching to a meat diet to heal their inflammatory bowel diseases is extremely difficult for them to do. The vegetarian diet promotes many different intestinal diseases. Vegetarians become so brainwashed that they can't chew meat and will gag when they attempt to swallow it—even though they are making an effort to eat it. Many women have been brainwashed to such an extreme that they can't eat animal fat. They claim it makes them gag, and they can't swallow the fat.

Universities are centers of higher brainwashing.

Many with religious restrictions gag and choke at the thought of eating pork, but they slurp it down unknowingly in a restaurant meal and say, "Wow, that was delicious." Yes, pork is delicious and a very healthy meat. We know that people who dislike animal fat are psychologically brainwashed because women in primitive societies who had never been brainwashed simply loved to eat raw animal fat. I grill lamb ribs just enough to sanitize the outside. The meat and fat inside are still raw. Overcooking will melt too much of the delicious fat. Raw lamb meat and fat are yummy—when you're are not brainwashed. Just ask any lion or wolf. By the way, the human digestive system is almost identical to those of the lion and wolf. We are designed with a high-acid stomach to digest meat and fat.

Chapter 6

Are Brainwashed Groups Mentally Disturbed?

Are left-wing liberal professors mentally nuts? "Yes," says an eminent psychiatrist who claims that those who possess a liberal ideology are suffering from a clinical mental disorder.

> "Based on strikingly irrational beliefs and emotions, modern liberals relentlessly undermine the most important principles on which our freedoms were founded," says Dr. Lyle Rossiter, author of the new book, *The Liberal Mind: The Psychological Causes of Political Madness*. "Like spoiled, angry children, they rebel against the normal responsibilities of adulthood and demand that a parental government meet their needs from cradle to grave."

Brainwashing by the Major Media Is Rampant

The media are major purveyors of brainwashing. Television programs spew unscientific nonsense to innocent children and adult masses who are already brainwashed. The brainwashing increases the power and control of the media as well as the income from special interest groups. Brainwashing techniques include shielding the population from the truth as well as dispensing false information.

Children and adults with childish mindsets easily become brainwashed by movies that depict animals with human characteristics and voices. The movie producers give the animals cute features and loveable childlike human personalities. The producers always portray mankind as mean and evil toward animals as they show the animals being harvested for our food and chimpanzees in human clothes talking and joking. They don't show chimpanzees as they really are. They don't show the chimp in an African National Park that attacked a woman, knocked her

Brainwashing of Society

to the ground, and tore her baby from her arms. They don't show the chimp carrying the baby off into the bush and eating the human infant alive. Don't believe the brainwashing that monkeys are vegetarians. These movies are brainwashing children into becoming vegetarians.

> **You have been brainwashed if you reject absolute truth or dislike eating animal fat.**

One common brainwashing story involves vitamin A and D toxicity. Animal-rights wackos have propagated lies about vitamin A and D toxicity from the consumption of cod liver oil. They know that many people will become brainwashed and avoid taking it. They don't want fishermen to catch the poor little cod, so vegetarians will tell you every lie under the sun in their effort to convince you to stop eating animal products. Other unknowledgeable people pass the false information on and on until nearly everyone accepts it. One teaspoon of Carlson's® Lemon Flavored Cod Liver Oil has about 20% RDA of vitamin A and 100% RDA of vitamin D, or 400 IU. I have taken two tablespoons—six times the recommended one teaspoon listed on the bottle—per day for several years. It has provided obvious health benefits without any sign of toxicity. A person would have to take 30 tablespoons per day for 6 months to reach the slightest level of toxicity of vitamin D. Polar bear liver can have a deadly level of vitamin D, and it killed early Arctic explorers after only one meal. I do not recommend polar bear liver on this diet.

Chapter 6

THE MERCK MANUAL*, *"Vitamin Deficiency and Toxicity."

"Vitamin D 1000 µg (40,000 IU)/day produces toxicity within 1 to 4 mo in infants; as little as 50 to 75 µg (2000 to 3000 IU)/day, if taken for years, can produce toxicity. In adults, taking 2500 µg (100,000 IU)/day for several months can produce toxicity."

You have been brainwashed if you think cod liver oil can be toxic.

A "global warming" brainwashing scare which showed Arctic polar bears on an ice floe looking very forlorn was presented on the Internet and TV on December 1, 2006. The report suggested the bears were trapped on the floating piece of ice and facing certain death. This brainwashing story runs several times a year on the Internet and TV. Actually, the bears are simply looking suspiciously at the humans nearby who are taking pictures. The story is nonsense and brainwashing. Polar bears can swim 100 miles in the open sea just for the fun of it, and they know where they are going.

Brainwashing of Society

The major story on December 1, 2006, was not about the polar bear trapped on the ice floe. It was about Chicago being trapped in the ice of a major winter blizzard. More than 400 flights were cancelled at Chicago's O'Hare International Airport, and 2.4 million people were without power. Ice was everywhere. Trees were collapsing under the ice load and taking down power lines. Schools were closed. The media chose the wrong day to try to brainwash people about the false global warming scare.

> **You have been brainwashed if you think global warming traps polar bears on a melting ice floe.**

Brainwashing by the Print Media Is Rampant

Book publishing is an interesting experience. The large book publishing companies do not accept inquiries—called *queries*—directly from the author. The query must first be sent to a literary agency that screens the incoming requests. Less than 5% of the queries submitted are accepted by agents. Some of the top authors had their first book rejected by many agents, who tend to follow the latest trends and topics and thereby extend the brainwashing. The agents submit manuscripts to publishing companies for the few authors they take under contract. The author is hopeful that at least one publishing company will offer to print the book. This method imposes a double censorship on the book. This must be why I rarely find a book to buy in the bookstore. It is easy to detect a brainwashing job by quickly browsing each book. The tables are stacked high with dozens of books by authors promoting a vegetarian diet in opposition to the Atkins' Diet. The attacks against Dr. Atkins are heated and insulting. Thousands of books are in the store, but very few are worth reading. Most books continue the brainwashing we see on TV because these are the in-vogue topics.

Chapter 6

Internet Free Speech Is under Attack by the Controlling Brainwashers

The obsessive compulsive controlling élite are having a hissy-fit over free speech on the Internet. They allowed this to happen because they thought they could corrupt the masses with smut, propaganda, Earth worship disguised as environmentalism, animal worship disguised as animal rights, and other brainwashing techniques. They totally missed the power that was being released to the individual—the insignificant man or woman in a basement or home office typing on a computer. The Internet allowed an ordinary citizen with very little money to blast messages around the world. China and other countries began blocking these messages immediately. All of the English-speaking countries are threatening to do the same.

The First Amendment to the US Constitution was written to allow free speech and freedom of the press. This hasn't stopped the US Government from using the Federal Communication Commission and a multitude of laws to control radio and television. Canada has gone a step lower by blocking religious broadcasting that proclaims the truth as written in the Bible.

The Second Amendment to the US Constitution, which provides freedom of firearm ownership and self defense, has been totally destroyed in some areas of the United States by the obsessive compulsive controlling élite. More restrictions are passed every year.

The United States is threatening to censor the Internet. They don't want to block the smut, Earth worship, animal worship, and scams. They want to block you from controlling your own life by reading absolute truth as presented in this book.

Brainwashing of Society

Brainwashing by Higher Education Occurs on a Massive Scale

Professors with common sense are blocked from entrance to universities by labeling them "extremely right-wing." Most—if not all—of the Ivy League universities were founded by God-loving Christians, but they are now teaching evolution and the Big Bang Theory. Elementary and secondary schools have done the same with teachers. Earth worship and animal worship in kindergarten and elementary schools are absolutely rampant.

> **You have been brainwashed if you believe Charles Darwin's Theory of Evolution.**

One hundred and fifty years ago, science was based on observable laws that repeatable experiments could prove. These laws govern many of the engineering and scientific fields to this day. Mechanical engineers depend on the truth of the laws of thermodynamics, heat transfer, fluid dynamics, and mechanisms. Electrical engineers faithfully employ laws concerning the flow of electricity and electromagnetic fields. Structural engineers apply laws concerning the stress and strain of materials in the design of giant structures. Something weird happened. Charles Darwin proposed the Theory of Evolution that had no basis in fact whatsoever. No one could prove it, and it violated observations in nature at every turn; but many people accepted it. The Theory of Evolution was born out of Darwin's self brainwashing.

The engineering fields would have dismissed Darwin's theory as just that—a worthless theory. Engineering is based on facts that can be proved by experiment—not on some pipedreams. Biologists were not as solidly based as engineers. They slowly began brainwashing themselves and others. Now the field of

Chapter 6

biology is one massive mess of nonsense. I cannot understand how universities can continue to list biology as a science. Music is closer to science than biology. The musical scales are purely scientific. Biologists prove that brainwashing breeds more brainwashing that can last for centuries.

Biology has become a massive mess of nonsense.

Physics, astrophysics, and cosmology were making great strides 100 years ago by using new telescopes that had ever-increasing power. Dr. Albert Einstein developed his *General Theory of Relativity* which showed that matter could be changed into energy. The atomic bomb proved him to be correct. The future for solid scientific achievements looked very bright.

Then brainwashing crept into the fields of astronomy, astrophysics, cosmology, and into the universities. Astronomer Percival Lowell (1855–1916) drew pictures of canals on Mars he said were constructed by Martians. He published three books on the subject, *Mars* (1895), *Mars and Its Canals* (1906), and *Mars As the Abode of Life* (1908). We know Mars does not have canals or alien creatures. How could Lowell have been so delusional? The canals were simply Lowell's imagination. Percival Lowell's brainwashing was influenced by Italian astronomer Giovanni Virginio Schiaparelli (March 14, 1835 - July 4, 1910) and the mistranslation of Schiaparelli's writings. Schiaparelli observed natural lines on Mars that he called by the Latin word, *canali*. This was mistranslated into English as *canals* instead of the correct word, *channels*. Lowell was misled by the English word into thinking they were constructed features, not natural as intended by the Latin word. Schiaparelli also described the flat basaltic plains on Mars as *maria*, the Latin word for *seas*. The

Brainwashing of Society

result was Lowell's vision of seas and canals on Mars as he peered through his low-resolution telescope.

Neither Schiaparelli nor Lowell could restrain his enthusiasm and imagination long enough to ponder the question, "Are the surfaces of Mars warm enough to allow flowing seas and water-filled canals?" The answer should have been obvious. Mars is much farther from the Sun than Earth, which would make finding liquid water on the surface of Mars much less likely. The ice caps at the poles of Mars contribute to the misconception by making it appear to be similar the Earth. Mars has frozen water, not flowing water.

The little false seeds planted by Schiaparelli and the mistranslation of the Latin grew in Lowell's mind into a monumental falsehood in which Martians were populating the planet and producing farm crops using irrigation canals. The path leading from unsubstantiated speculation of seas and canals on Mars to people living there and farming has been surpassed today by the expanding myth called the *Big Bang Theory* and the expanding myth that carbohydrates are healthy and necessary in the diet.

Another monumental stumble in science occurred when Dr. Edwin Hubble concluded that the light coming from distant galaxies was shifted to the red spectrum because the galaxies were all moving away from Earth at an accelerating rate. Tracing the direction and velocity of the galaxies produced the concept that the universe existed at a point in the finite past. Boom! The Big Bang Theory was born. Dr. Hubble should have given it more thought. The most likely cause for the red shift of light is the slowing of the speed of the light—not the acceleration of the distant galaxies. The massive brainwashing was off and running. Astronomers, astrophysicists, and cosmologists could now cite a

Chapter 6

theory for the creation of the universe that excluded Intelligent Design. It didn't matter that the theory violated many of the existing laws of science. Brainwashed people continue to accept both of them.

> **You have been brainwashed if you believe the Big Bang Theory. Read the next chapter to see why.**

Preschool cartoons brainwash today's children into believing the bogus Theory of Evolution and Big Bang Theory. The elementary school system continues to refine the brainwashing. The poor children are helpless and unable to reject these unproven theories. They grow up to become brainwashed Ph.D.'s who teach the same nonsense to their students.

Biologists, astronomers, astrophysicists, and cosmologists were on solid scientific footings 150 years ago. Now they have drifted into space on theories devoid of scientific validation. The same can be said of modern nutritional professionals.

Testimony from a High School Student Who Escaped the Attempted Brainwashing

"I totally agree with you that evolution is wrong. When I was a kid in elementary school I used to believe in evolution, but when I became a born-again Christian I was like "how can there be evolution?" Now I spend a lot of my time researching about it, even though it's really for a research paper that my Mom wants me to do for my home school paper (I'm in high school). I can't wait to grow up and become a scientist. Sometimes I feel that God put me on earth to help prove evolution wrong, but who knows?

Brainwashing of Society

Thanks for the info., and I really like the stuff about the obsessive control freaks and all the brainwash. I sort of always thought that I had been brainwashed as a kid..."
Thanks,
Michelle

The topics of diet and health are awash with brainwashing from the news media, universities, books, and questionable studies. The following study is a good example.

"High IQ link to being vegetarian" *BBC News*- December 15, 2006.

> "Intelligent children are more likely to become vegetarians later in life, a study says. A Southampton University team found those who were vegetarian by 30 had recorded five IQ points more on average at the age of 10."

The report gives the false impression that the vegetarian diet is best because some higher-IQ children became vegetarians. The report has several serious flaws. First, the higher-IQ children gained their smarts on a meat diet. They did not become vegetarians until later. Very few adults today were fed a vegetarian diet as infants. Second, the higher-IQ children are more likely to attend and graduate from college because of their higher IQ. The college education simply brainwashed them into believing the vegetarian diet is healthier—which it isn't. The lower-IQ group did not become vegetarians because they were not brainwashed on college campuses. The study simply proves that universities are brainwashing students.

Chapter 6

Brainwashing by Mass Marketing Groups and Salesmen

People in all English-speaking countries are brainwashed from birth by mass marketing techniques and salesmen. These people have failed to develop mental thinking based on investigation, logic, and discernment of the truth. Rather than using sound logic techniques, they respond to the best sales pitch. People believe the information presented by dynamic speakers, articles, books, or TV shows that best appeals to their feelings. Most people make choices by feelings, not by truth or logic. These people are easily brainwashed.

The message about diet, nutrition, and health fed to the masses for the last 100 years is a good example. Food manufacturing companies teach people that they must eat carbohydrates for energy. Other salesmen refine the message somewhat to say that some carbohydrates are good and others are bad. The current epidemics of obesity, diabetes, cancer, and heart disease in all English-speaking countries are the result of this brainwashing. The entire world has been brainwashed into eating carbohydrates.

You have been brainwashed by a corn oil salesman if you believe saturated fats cause heart disease.

Another good example concerns saturated fats in the diet. The average brainwashed individual believes eating saturated fats is detrimental to health and leads to heart disease and cancer. This false teaching was propagated nearly 100 years ago by food manufacturing companies that wanted to sell their omega-6 vegetable oils for cooking to compete with animal fats. The brainwashing worked on the masses. Saturated fats are scientifically very healthy, but the omega-6 vegetable oils are unhealthy.

Beef is one of the healthiest foods anyone could eat. Beef is naturally loaded with essential protein, fats, vitamins, minerals, enzymes, and a multitude of other dietary supplements needed for optimum health. All of the negative claims against eating red meat are false, and the negative studies are fraudulent. A diet consisting of 100% red meat would provide wonderful health. The claims that supermarket beef contains harmful antibiotics, toxins, hormones, pesticides, bad fats, and diseases are simply false.

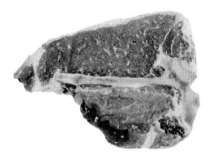

The above examples of brainwashing concern diet, nutrition, and health. Similar brainwashing has occurred in any and all subjects you could name. When the truth is presented to the brainwashed masses, they simply cannot comprehend it. They have a conflict between truth and what they have been brainwashed into believing. Their brains simply lock up.

Brainwashing of the Masses Can Last for Centuries

People once thought the Earth was flat and was the center of the universe. The myths lasted for centuries. We now know this was silly brainwashing.

For centuries doctors have bragged that the practice of medicine is an art, not a science. Perhaps we should applaud them for their honesty because medicine did not begin as a science. Medicine began as witchcraft and has a history of sidestepping scientific facts at every opportunity.

Chapter 6

Doctors' finest achievements in medicine are their abilities to perform surgery on delicate tissues of the body, and they should be highly commended for this skill.

Doctors have been severely brainwashed by their professional societies and medical schools. They continue to believe that heart disease, cancer, and diabetes are caused by eating red meat, saturated fat, and cholesterol. Study after study is performed which proves that saturated fats do not cause heart disease, but not a glance is received from the majority of the medical community. Doctors rejected the proven theory that bacteria caused flesh infections for 200 years before they stubbornly relented. Imagine, the doctors operated on people with dirty hands and instruments for 200 years after the discovery of bacteria. The idea that medicine is an art and not a science prevails.

Most gastroenterologists continue to insist that ulcerative colitis and Crohn's disease have nothing to do with diet. By taking such a stand, these specialists never bother to test or research the effects of diet on these inflammatory bowel diseases. Doctors in alternative medicine have dismal results as well because they embrace unscientific theories, such as acupuncture, homeopathy, Chinese medicine, Hindu, and Buddhist medicine—all of which border on being nothing more than witchcraft.

You have been brainwashed if you believe in acupuncture, homeopathy, or Chinese medicine.

Doctors continue to claim that sugar and carbohydrates do not cause diabetes. Some have even taught that saturated fats cause diabetes. This boggles the minds of people who base the causes of diseases on scientific facts. Doctors simply blame the patient

Brainwashing of Society

when the high-carbohydrate diet recommended by the American Heart Association and the American Diabetes Association doesn't make him better. It is his fault because he either didn't get enough exercise or he has the heart disease and diabetes genes in his family. These false excuses have become the catchall escape routes for the failures of modern medicine.

Brainwashed Victims are Simply Ignoring Scientific Facts

The professional nutritionists are the most brainwashed and unscientific group of the lot. Science and facts mean nothing to them. They go even farther by performing fraudulent studies that they claim are scientific. Other scientists cannot duplicate their results, but the nutritionists stand firm in their positions and defend their brainwashed ideas. Engineers never get themselves into this type of quagmire. Engineers have developed laws and formulas that cannot be disputed. Theories, research tests, and field applications validate the formulas they use. This is why planes fly, bridges stand, and power generating stations keep producing electricity. Engineering failures occur because someone makes an error in the application of the theory—not because the theory is wrong. Professional nutritionists have given us unrelenting epidemics of heart disease, diabetes, and cancer because of their unscientific propaganda.

> **You have been brainwashed if you believe the vegetarian diet is healthy.**

The vegetarian diet has been presented as a healthier way of eating, but nothing could be farther from the truth. Vegetarians are getting sick with intestinal diseases at alarming rates, especially young adult females, who are more inclined to become vegetarian than males. Monitoring Internet vegetarian

Chapter 6

nutritional message boards reveals that a high percentage has inflammatory bowel diseases such as ulcerative colitis and Crohn's disease. Their vegetarian diets do not make them well but continue to make their poor health conditions worse. This is not surprising considering the fact that their diet is very high in carbohydrates.

Brainwashing Includes the Acceptance of Conflicting Information

The fact that socialism and communism are unsuccessful economic systems doesn't seem to bother the brainwashed in the slightest. College students and politicians can be heard espousing socialistic economic dogma even though recent years have proved the disastrous results produced by these economic systems. The economy of the Union of Soviet Socialist Republics (USSR) collapsed within a few decades under the communistic system. The Chinese people have become the world's slaves and produce goods on meager wages. The people of Cuba have a living standard that is lower than most third-world countries.

> **You have been brainwashed if you believe government social programs can eliminate poverty.**

Brainwashed scientists can hold several conflicting theories and believe that all are true—even though that's impossible. They talk with complete confidence about one theory and later talk about the other theory with equal enthusiasm. The Big Bang Theory is a good example. College professors talk about matter expanding faster than the speed of light during the early Big Bang period but later teach Dr. Einstein's theory that matter cannot travel faster than the speed of light. These kinds of

Brainwashing of Society

conflicts would normally be classified as mental illness—except when they involve brainwashed teachers.

If the person installing tires on your car told you they needed 40 psi of air pressure but later told you that driving without any air pressure in the tires was also OK, you would certainly say, "This guy is nuts." On the other hand, a college professor will tell his students that for millions of generations the ancient bird carried heavy wings that were too small for flight. Later the professor says that natural selection gave the bird wings large enough to fly. The students just sit there—brainwashed. Jump up out of your seat! Say something! The Theory of Natural Selection dictates that generations would produce smaller and smaller, useless wings, not larger wings because larger wings are a handicap to a flightless bird. The Theory of Natural Selection proves evolution to be wrong. The professor simply ignores the conflict in logic.

Are Students Being Educated or Brainwashed?

Chapter 6

Help me! I can't fly!
My wings are too small, and my head is too big.
I have been brainwashed.

You have been brainwashed if you believe birds evolved larger and larger wings over millions of generations until they could finally fly.

Brainwashing Is Done on the Basis of Feelings, Not Logic

The most effective method for brainwashing someone is to appeal to his feelings. Logic is rarely used in brainwashing. The brainwashed individuals must be deprogrammed by using logic to prove that the ideas in their minds are in conflict. All of the beliefs of the brainwashed cannot be true. Some must be discarded. Little by little the brainwashing is reversed as the person learns and accepts the truth.

Brainwashing of Society

Environmental and animal rights activists are good examples of brainwashing. They can be seen swooping down in a helicopter, firing a tranquilizing dart into an animal, placing a tracking collar around its neck, and keeping it under constant surveillance. Soon the animal is on the Endangered Species List. This is nuts. Animals that provide great steaks for a cookout are never on the Endangered Species List. The best thing that could have ever happened to the California condors would have been to advertise that they are much better to eat then turkeys. Soon farmers everywhere would be raising condors in the same way they raise ostriches.

Brainwashers Have an Obsessive Compulsive Controlling Disorder (OCCD)

Those who attempt to brainwash others are suffering from obsessive compulsive controlling disorder (OCCD). In his private life, his mate and children are miserable. In the business world, the obsessive compulsive controller advances to positions of leadership by assumed authority. He must review and approve every detail. He does not delegate well. His subordinates sit around doing nothing while he flounders in a self-imposed overload. The subordinates keep silent about an impending crisis and laugh when he crashes. Controllers are empire builders with vast organizations of unproductive subordinates. The obsessive compulsive controlling manager can literally destroy a company.

People with obsessive compulsive controlling disorder have over-inflated egos, and they go to great lengths in their attempts to force their opinions on others. They are constantly frustrated in life because everyone ignores them. The only stable relationships they have are with those suffering from compulsive submissive disorder.

Chapter 6

The person with OCCD is usually wrong about the things he is trying to get others to do. He doesn't take his own advice, and he never uses absolute truth as the basis for his thinking. Politicians who suffer from OCCD attempt to force their ideas on everyone else and make us pay for it with higher taxes. They seek to create utopia but it always ends in a mess.

Brainwashed Individuals Can Have Obsessive Compulsive Debating Disorder (OCDD)

Many of those who become brainwashed try to push their false ideas onto everyone around them. They are constantly debating these false ideas with friends, relatives, coworkers, and neighbors. They write to websites and object to just about everything in their compulsion to start a debate. Many times they ask the same question again and again by slightly rewording it. When they are presented with scientific facts, they simply ignore them and go on as if they are still right. They never come to the knowledge of the truth.

Normal people will generally answer a question or express an opinion in response to the debater's question, but the debater must eventually be ignored because no answer is satisfactory. This frustrates the individual with the debating mental disorder, and he loses control and gets mad when no one will engage in a debate. The debating obsession is a form of assault and control.

The person with obsessive compulsive debating disorder can find happiness in meeting another person with the same mental illness. They hammer away at each other, but neither is listening to a word that the other person is saying.

Absolute Truth Exposed

Volume 1

Chapter 7

Absolute Scientific Proof the Big Bang and Redshift Light Theories Are Wrong

Chapter 7

This chapter will provide scientific proof that the Big Bang Theory is wrong. The Big Bang, Redshift Light, and Dark Energy Theories have many major flaws, errors, and problems.

The Big Bang and Redshift Light Theories are not laws of science, physics, or astrophysics. A scientific law must be 100% correct. Failure to meet only one challenge proves the law is wrong. This book will prove that The Big Bang and Redshift Theories fail many challenges—not simply one. The Big Bang Theory will never become a law of science because it is wrought with errors. The light from distant galaxies does have a redshift, but the Redshift Light Theory is also false. This is why they are called *theories*, not *laws*. The scientific facts below will show why light from distant galaxies has a red shift. The Big Bang and Redshift Light Theories will never become laws of science; no one can prove them to be scientific laws because they are false concepts. The Big Bang and Redshift Light Theories are perpetuated by brainwashing the general public and high school and university students. The following is the forbidden science question on campus.

Does the speed of light decay as it travels through space?

It actually does, but due to his advanced brainwashing, your professor will tell you it does not.

Scientific support for the Big Bang Theory is very weak. In fact, there is none. The theory is pure conjecture based only on the Redshift Light Theory as proposed by Dr. Edwin Hubble. A considerable amount of scientific evidence harshly opposes the Big Bang Theory, but supporters simply ignore this evidence.

Proof the Big Bang Theory Is Wrong

Big Bang Flaw No. 1

The Expanding Universe Theory

The Big Bang Theory violates the basic foundation of science that is Dr. Albert Einstein's *Special Relativity*.

$$E = M * C^2$$

Dr. Albert Einstein developed this formula that gives the relationship between energy (**E**), mass (**M**), and the speed of light (**C**). The theory has been proven in numerous ways. It is the theory behind the atomic bombs for both fusion and fission reactions. Dr. Einstein conducted experiments to prove his theory. It stands without a challenge.

The Big Bang Theory states that the universe was once a dense, hot body of energy that exploded and has been expanding since the beginning, some 10 to 20 billion years ago. Dr. Einstein's formula shows that material cannot exceed the speed of light as it expands. Mass becomes infinite as it approaches the speed of light. Increasing the energy in the initial "bang" cannot increase the speed beyond the speed of light. The mass will increase proportionately as the initial energy is increased. The mass increases to approach infinity as the energy level approaches infinity. The Big Bang Theory violates this scientifically proven law discovered by Dr. Einstein. This scientific law alone should have kept the Big Bang Theory in the closet forever, but people simply ignore reality and the truth.

The speed of light seems fast to us, but it is only a tiny fraction of the speed needed for the expanding universe in the Big Bang Theory. The speed of light is 186,000 miles per second—300,000 kilometers per second. This is a snail's pace in cosmic terms. A

Chapter 7

cosmic distance of 186,000 miles is negligible. One second is like an eternity in Big Bang terms.

Massive Explosion on Our Sun

The Big Bang Theory has a gravity problem. A speed near the speed of light is inadequate for mass to escape the tremendous gravitational field at time zero in the Big Bang Theory. The matter would simply decelerate sharply and collapse back onto itself again. The speed of light is much less than the escape velocity needed for matter during the Big Bang explosion. The Big Bang would simply give a big burp and collapse back again; the energy would be changed back into mass. This condition can be witnessed daily by observing the Sun. Giant explosions called *sun spots* blast material hundreds of thousands of miles (km) into space only to decelerate and collapse back onto the Sun again.

Proof the Big Bang Theory Is Wrong

Radiation and light escape from the sun but heavier matter cannot.

> **Astrophysicists made a big mistake by treating the Big Bang Theory as Law. Now they can't correct the blunder.**

One cosmologist has suggested that the material forming our universe blasted out from the "Big Bang" at a speed greater than the speed of light. At least this guy can see the speed of light as a barrier to the *Expanding Universe Theory*. However, having matter move at a greater speed than the speed of light is totally devoid of any theoretical formulas, violates Dr. Einstein's *Special Relativity*, and lacks astronomical observations to prove it. Nothing in the universe has been shown to travel at a speed greater than the speed of light except some forms of radiation. These scientists are approaching the problem backward. Before the theory is accepted, the scientific formulas and laws must be developed to support the theory. Instead, the scientific community has accepted the Big Bang Theory without any science or observations to support it. Now they are simply throwing out any wild idea imaginable, hoping to keep the Big Bang Theory from collapsing.

The Big Bang Theory is based on energy being converted to matter. Scientists are using particle accelerators in an attempt to prove this theory, but the success has been very limited. Scientists have been unable to create even one atom of hydrogen. The atomic bomb turns matter into energy, but this process is highly irreversible. The Big Bang Theory is an atomic bomb in reverse. The scientific basis is lacking. The energy from an atomic bomb does not revert back into matter. The claim that all matter in the universe was created from energy is a myth.

Chapter 7

There is a big difference between mass and matter. A rocket motor behind an asteroid will accelerate it across the universe, and the energy required will increase as the asteroid approaches the speed of light. The mass will increase, but this does not mean the matter will increase. The number of atoms in the asteroid will remain unchanged. According to Dr. Einstein's formula, the resistance to further acceleration will appear as an increase in mass. Matter is not created by the energy imparted to the asteroid. The Big Bang could not create matter—galaxies, stars, and planets as shown in the cluster of galaxies below.

The Big Bang Theory does not resolve the big question, "Where did we come from?" The bigger question would be, "Where did the initial energy before the Big Bang come from?" Supporters of the Big Bang Theory treat the initial bang as the beginning of all things. They are ignoring the obvious questions instead of giving an answer. I will give the answer for the cluster of galaxies.

Artist's Illustration of the Big Bang

Proof the Big Bang Theory Is Wrong

Big Bang Flaw No. 2

The Redshift Light Theory

The Redshift Light Theory is in big trouble. Perhaps this should be listed as Big Bang Flaw No. 1 because it gave birth to the Big Bang Theory. The Redshift Light Theory is the very foundation for the Big Bang Theory. The theory, first formulated by Dr. Edwin Hubble and his assistant, Milton Humason, in 1929, soon became known as the *Hubble Redshift Theory*. Dr. Hubble's conclusion was immediately and almost unanimously accepted by astronomical scientists without any scientific proof. The Redshift Light Theory quickly led to the Big Bang Theory with virtually no scientific proof.

The concept of a universe expanding from a single source at a real point of time in the past was developed because Dr. Hubble discovered the red light shift. He discovered that very distant galaxy clusters were emitting light with a redshift and theorized that the distant galaxies were moving away from us because the light emitted had a shift to a longer wave length caused by the velocity. This is called the *Doppler effect* in sound waves. White light emitted from an object that is moving away at a high speed appears in the red spectrum. The light appears in the violet or blue spectrum when the object is moving toward the viewer. The concept is simple—perhaps too simple. The Redshift Light Theory has our little planet at the center of the universe because nearly all galaxies are moving away from us. This should have been the first clue that his theory was just too simplistic. The probability that the Earth lies at the center of the universe is a ridiculous myth that never seems to die.

Chapter 7

Parents correctly tell their children they are not the center of the universe.

Dr. Hubble also noticed that more distant galaxies had a greater red light shift than closer galaxies. He postulated that the more distant galaxies were moving away faster (accelerating) at a higher velocity than the closer galaxies and claimed the higher acceleration was directly proportional to the distance away from Earth. However, modern scientists have seen the big problem with Dr. Hubble's theory. A galaxy does not have a propulsion system to make it accelerate because the obvious gravitational forces from other galaxies would tend to collapse the universe. University professors and all other scientists who have supported the Big Bang and Redshift Light theories have been forced to devise an unseen, unproven, unknown force, or unknown object in the universe that would resolve all of the missing science in their theories. This force or object had to be black, emit no X-rays, emit no radiation, emit nothing, and have no effect on light—because no such object can be seen or detected in the universe. They called this hypothetical scientific scapegoat *Dark Energy*. Cool! The entire scientific community now depends on the fictional dark energy as the foundation for The Big Bang and Redshift Light theories. Dark energy fills in all of the gaps where no science exists. This is not a little fudge factor or tweak to make theory agree with the physical observations of the universe. This is the major foundation for cosmology. The dark energy must be many times larger than all of the known matter in the universe to make the Redshift Light Theory plausible. The description of the mythological dark energy is even more unbelievable and bizarre. Dark energy must exert a negative force on visible matter. In other words, it must behave like negative gravity. It pushes galaxies around.

Proof the Big Bang Theory Is Wrong

Therefore, scientists have termed it *gravitational repulsion*. Scientists also use terms such as *dark matter, dark flow, exotic matter, negative mass, and anti-matter*. Dark objects are real—like planets in orbit around distant stars. This is not what they are talking about. Another big scientific void is created by Dark Energy. There are no scientific proofs that gravitational repulsion exists or is even possible. In fact, there is not even a nitwit theory. No proof. No measurement. No description. No nothing. Just magical Dark Energy. The Dark Energy fairy is pushing the universe apart. No! We should also be aware that the Dark Energy Theory is a blatant violation of Dr. Albert Einstein's *Special Relativity* and his *Theory of General Relativity*. Dr. Einstein rejected such thoughts as gravitational repulsion or *anti-matter*. I must give a warning to university students. Don't let them brainwash you with the Dark Energy nonsense!

Hubble's conclusion has never been seriously challenged until now. This challenge is not based on a religious alternative as many would expect, although the Biblical description of creation could fit into a big bang scenario. Many religious people consider the first verse of the Bible (Genesis Chapter 1, Verse 1) to be a description of the Big Bang and Jeremiah 10:12 to be an expanding universe.

> **Genesis 1:1** "In the beginning God created the Heavens and the Earth."
>
> **Jeremiah 10:12** "He has made the earth by His power, He has established the world by His wisdom, And has stretched out the heavens at His discretion."

I will explain the absolutely true scientific reason for the redshift in the light from distant galaxies. The Doppler effect assumed by Dr. Hubble is not the reason light from distant galaxies has a

Chapter 7

redshift. The scientific truth destroys the Big Bang Theory as well.

The idea that the speed of light is a constant is nebulous at best—a little humor there. In our small little world, the speed of light appears to be constant only because we cannot measure the change over a long distance—I mean it couldn't be measured until now. This problem is analogous to saying the speed of a bullet fired from a gun is constant. It appears to be constant if you measure the bullet velocity one meter from the end of the barrel and measure it again two meters from the barrel. However, the truth is revealed if you measure the velocity 1000 meters from the barrel. The air resistance and gravity force slow the bullet as predicted by the laws of science.

What about light? Light has mass and the bullet has mass. The fact that light has mass was proved and measured many years ago. Light is affected by the gravity produced by other masses; it bends as it travels past distant planets, stars, and galaxies. Certainly the numerous masses in the universe can slow or accelerate the light by gravitational attraction. White light emitted by a stationary object will appear as redshift light if the velocity is slowed for any reason. So what made Dr. Hubble think that the light speed was not affected by gravitational fields in the universe? I suppose he didn't stop to consider all of the alternatives.

$$F = M * A$$

Light traveling through the universe is slowed or accelerated by the gravitational fields of stars and galaxies. Light conforms to the proven formula **F = M * A**, in which force **(F)** is the product of mass **(M)** times acceleration **(A)**. Cosmologists calculate that 90% of the mass in the universe cannot be seen—this is probably

Proof the Big Bang Theory Is Wrong

wrong as well. The visible stars and galaxies only comprise 10% of the mass of the universe. The black mass is there and has an effect on light passing nearby.

Abell 2218: A Galaxy Cluster Lens
Credit: Andrew Fruchter (STScI) et al., WFPC2, HST,
NASA Digitally reprocessed: Al Kelly

NASA explains how this picture Abell 2218 proves that light bends as it travels through spacetime. If spacetime can bend light, it can also change the speed of light.

> "Explanation: What are those strange filaments? Background galaxies. Gravity can bend light, allowing huge clusters of galaxies to act as telescopes, and distorting images of background galaxies into elongated strands. Almost all of the bright objects in this Hubble Space Telescope image are galaxies in the cluster known as Abell 2218. The cluster is so massive and so compact that its gravity bends and focuses the light from galaxies that lie behind it. As a result, multiple images of these

Chapter 7

background galaxies are distorted into long faint arcs -- a simple lensing effect analogous to viewing distant street lamps through a glass of wine. The cluster of galaxies Abell 2218 is itself about three billion light-years away in the northern constellation of the Dragon (Draco). The power of this massive cluster telescope has allowed astronomers to detect a galaxy at the distant redshift of 5.58."

All stationary galaxies would appear to be traveling away from Earth in every direction if the light from those galaxies were slowed as it traveled through space. The light from a more distant stationary galaxy would be slowed more because of the greater distance from Earth. This is exactly what the redshift measurements show. A reduction in the speed of light as it travels through the universe is my scientific discovery that destroys the Big Bang Theory. Absolute truth has been exposed in this book.

Scientists claim that antimatter exists, but does it? How does it affect light? Where is the antimatter? Antimatter would erase or cancel light, but we see no evidence of that. Astronomers have never seen a galaxy or star simply disappear after being swallowed by antimatter. From the scientific evidence and the lack thereof, I conclude that antimatter is another figment of the science-fiction imagination.

The vast openness of the universe appears to be empty, but it may contain clouds of undiscovered particles smaller than an electron which slow the speed of light. The science against the Redshift Light Theory is immense, and the Big Bang Theory is in serious jeopardy. Remember, a law of science must withstand each and every assault and test. The Big Bang Theory cannot do that.

Proof the Big Bang Theory Is Wrong

Rieske Spacetime Drag Coefficient for the Speed of Light

In his *General Theory of Relativity,* Dr. Einstein proposed that mass causes a wrap in time and space that he called *spacetime.* I propose a new theory that the speed of light slows as it travels through the universe, and I am calling this new discovery the *Rieske Spacetime Drag Coefficient for the Speed of Light (RSDC).* This proclamation was first made worldwide on an Internet website on November 8, 2005.

In the past, science has incorrectly assumed that the speed of light is constant, but I will show why the velocity of light decays as it travels through space. The resistance imparted by spacetime is the primary reason the speed of light decreases. The reduction in the speed of light through space has three components, all of which contribute to the decline in the speed of light and are in complete agreement with Albert Einstein's *General Theory of Relativity*.

- Light photons have mass that deflects the essence of spacetime and creates a gravity wave that imparts resistance to the movement.
- Light traveling across spacetime changes time. The elapsed time as measured by an astronomer's watch is longer than the time as measured by an imaginary watch attached to the light photons.
- Light photons encounter other mass throughout the universe which is not an empty vacuum as has been assumed in the past. Ninety percent of the mass of the universe is believed to be matter that does not give off light. We cannot see, identify, or measure this hidden mass, but it certainly contributes to the decline in the speed of light as it travels across the vast universe.

The amount of slowing of the speed of light as it crosses the universe is the *Rieske Spacetime Drag Coefficient for the Speed of*

Chapter 7

Light. I will calculate this coefficient based on existing scientific data.

A large mass will produce a large warp in spacetime. Dr. Einstein proposed that by traveling fast in space and returning to the point of origin, one will experience a smaller amount of elapsed time than that of a person who remained at the point of origin. The space traveler is a little younger than he would have been had he not traveled at a high velocity in space. This phenomenon is known as *time dilation* and is an irreversible time resistance that has been proven by scientific experiment. However, Dr. Einstein failed to propose that the spacetime warp also causes a decrease in the velocity of matter traveling through space. He taught that the speed of light was a finite, constant value. I propose that the velocity of the mass is subjected to resistance by the warp of space and time. The reduction in the velocity of mass in spacetime also increases entropy and is irreversible. This resistance can be thought of as similar to a ball rolling on a surface. The warp or deflection of the surface and ball will cause a resistance that is proportional to the amount of warp. Engineers use this factor daily as the rolling friction resistance that increases entropy and is irreversible. Likewise, a mass traveling in space is not resistance free. The mass is constantly pushing against the spacetime warp.

> ### Rieske Spacetime Drag Coefficient for the Speed of Light absolutely destroys the Redshift and Big Bang Theories.

Light has a very small mass which creates a very small spacetime warp. However, the spacetime warp created by light is not zero. Light has a very high speed, and we measure the distance traveled in millions of light-years. The velocity of light emitted

Proof the Big Bang Theory Is Wrong

from a distant object in the universe slowly decreases due to the *Rieske Spacetime Drag Coefficient*. The speed of light emitted from the most distant galaxies decreases the most because the *Rieske Spacetime Drag Coefficient* affects the light over the longer distance—not because the galaxy has a higher expanding velocity as Dr. Hubble taught. The change in the speed of light caused by the spacetime resistance appears as a redshift in the visual spectrum even though the distant galaxy has no differential velocity. Therefore, the more distance galaxies are expected to have the greatest redshift in the velocity of the light because light travels farther and velocity decreases more. This is exactly what Dr. Hubble observed. My theory does not suggest that the Earth is at the center of the universe as does the Redshift Light Theory. The redshift occurs for galaxies that are also static in space. The universe does not have a center because it does not have an outer limit. The universe could be infinite. Our vision is limited, but the universe may not be. Future telescopes that are much stronger than the Hubble Space Telescope will find galaxies in space extending to the reaches of present observation and beyond. Dr. Hubble observed the *Rieske Spacetime Drag Coefficient* in action, but he gave the wrong explanation for the phenomenon. Dr. Hubble incorrectly assumed the galaxy velocity accounted for all of the redshift. He failed to recognize that a drop in the speed of light due to spacetime warp resistance was also occurring and accounted for the majority of the redshift. The universe could actually be in a state of collapse while still exhibiting a redshift in the light due to the *Rieske Spacetime Drag Coefficient*. This is more likely our current state. The universe is static or slowly collapsing. The theory that the universe is collapsing has a strong foundation because that is what we would expect due to the gravitational attraction of the total mass of the universe, especially if the universe is not infinite.

Chapter 7

The resistance or friction on a mass moving in spacetime also has an effect on time. *The Rieske Spacetime Drag Coefficient* causes time as experienced by the mass to slow relative to other mass that is not moving in spacetime. This phenomenon has already been proven by sending an accurate clock into space and back. The velocity of the clock in spacetime resulted in less elapsed time than the clock remaining on Earth. The velocity in spacetime is proportional to the rate of reduction in time. This time reduction is not a reversible process, and resistance or friction of spacetime results in a reduction in velocity as well as time. Travel in the vacuum of space is not a frictionless motion; thus the universe is slowing, not accelerating as stated by Dr. Hubble. The slowing of the universe results in a slowly collapsing universe.

The speed of light is usually expressed in units such as meters per second. Time is the denominator. Distance is the numerator. Light traveling from a distant galaxy arrives in the astronomer's telescope after a certain amount of elapsed time. A clock attached to the light photon would give a much shorter elapsed time than a clock on Earth—as proved by NASA. This means that the denominator based on Earth time is larger than on the light clock. The light may have been traveling at the speed of light based on the light clock, but the speed of light based on Earth's elapsed time is slower. In other words, the travel time on the Earth clock is longer, which means the light was traveling slower. The slower speed would result in a redshift in the light just as Dr. Hubble observed. The redshift is caused by a difference in time and spacetime drag, not by the distant galaxy moving away from us at a high rate of speed. Dr. Hubble was wrong.

Imagine a bowling ball rolling on the hardwood floor of a bowling alley. The resistance is very small, and the ball does not

Proof the Big Bang Theory Is Wrong

lose much speed. Now imagine the bowling ball on a rubber floor on which the weight of the ball deflects the rubber, creating a wave in front of the ball. The ball is constantly attempting to climb the wave and thereby creates a resistance that slows the velocity of the ball. The travel of a mass through spacetime is similar to the bowling ball on a rubber floor. The mass in space deflects spacetime (spacetime warp) and creates a resistance (gravitational wave). Objects with mass do not freely travel through the voids of space as is claimed in science books and taught in college classes. The resistance (gravitational wave) causes the mass velocity to decay. This condition was predicted by Einstein's *General Theory of Relativity* and has been proven by real observations. The binary pulsar PSR 1913+16, discovered in 1974, is a neutron star and a pulsar in orbit around each other. The pulsar allows extremely accurate measurements of orbital time. Frequent measurements have confirmed that the travel of the pulsar through spacetime has caused the orbital time to decrease 20 seconds in 20 years. These measurements scientifically prove beyond any doubt that spacetime warp causes the velocity of a mass to decay or decrease.

Absolute Truth Has Been Exposed.

The deflection or warp of spacetime by a light particle is indeed small, and the gravitational wave (resistance) is indeed small. But the small deflection (warp) of spacetime affects the light particle because the mass of light is also small. The velocity of the light decays over vast distances of the universe in the same manner and for the same reason as does the constant orbital time decay rate of binary pulsar PSR 1913+16. The following is a quote from NASA:

Chapter 7

"In 1967, the first radio pulsar was discovered by Jocelyn Bell and Anthony Hewish. Pulsars were quickly identified as neutron stars, the incredibly compressed remnants of the supernova explosions of main-sequence stars. In 1974, Russel Hulse and Joseph Taylor discovered the first binary pulsar, PSR 1913+16, a system of two neutron stars (only one emitting as a pulsar) with an orbital period of eight hours. As shown in the figure, the orbit of the binary decays exactly as expected for system of accelerated masses that lose energy to gravitational waves."

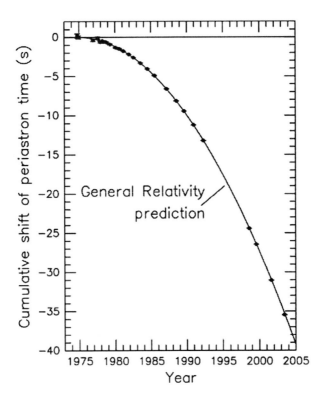

Credit Image: Weisberg and Taylor 2004

Proof the Big Bang Theory Is Wrong

"Hubble Reaches 'Undiscovered Country' of Most Distant Primeval Galaxies"

"*ScienceDaily* (Jan. 5, 2010) — NASA's Hubble Space Telescope has broken the distance limit for galaxies and uncovered a primordial population of compact and ultra-blue galaxies that have never been seen before."

"The most remote galaxies discovered in the universe are estimated by astrophysicists to have been 13 billion light years away at the time the light left the galaxies just 600 million to 800 million years after the Big Bang. These galaxies were discovered by the Hubble Space Telescope."

"The deeper Hubble looks into space, the farther back in time it looks, because light takes billions of years to cross the observable universe. This makes Hubble a powerful "time machine" that allows astronomers to see galaxies as they were 13 billion years ago, just 600 million to 800 million years after the Big Bang."

This report raises several disturbing questions. How could the newly-discovered galaxies have moved from the center of the Big Bang to a position 13 billion light years away in only 600 million to 800 million years? Remember, astrophysicists tell us that we are at or near the center of the Big Bang. This report infers that the galaxies would have had to travel 18 times the speed of light—a ridiculous concept. Outlandish statements like this appear in astrophysicists' reports almost daily without the blink of an eye, snicker, or challenge.

More disturbing is the statement that the galaxies were "ultra-blue." According to the Redshift Light Theory, these galaxies are traveling **toward us** at a very high rate of speed. Were they blown out by the Big Bang to a position 13 billion light years

Chapter 7

away in only 600 million years but quickly reversed course to be speeding back toward the center of the Big Bang with a velocity that produces ultra-blueshift light? Galaxies cannot reverse direction. This newly-discovered group of blueshift galaxies proves the Big Bang Theory is a joke. The newly-discovered blueshift galaxies indicate the universe is collapsing, not expanding. The Big Bang Theory is dripping with contradictions. The Hubble Space Telescope has dealt the Big Bang Theory another fatal blow.

> **The Big Bang Theory was based on distant galaxies emitting redshift light. Now we have found more distant galaxies emitting ultra-blueshift light. Uh-oh! Big Bang is in trouble again.**

The velocity of light is expressed as $c = \lambda * f$, where λ equals the wavelength and f equals the frequency. Time changes when a mass travels through space. Light has mass, and the frequency is a function of time. The result is a change in the speed of light.

I will calculate the *Rieske Spacetime Drag Coefficient (RSDC)* by accurately determining the decrease in the speed of light (decrease in frequency) as it travels across vast distances of spacetime.

I will accept the distance to remote redshift galaxies to be 12 billion light years based on the average redshift of light from 2004 observations and earlier. My calculations assume a static universe rather than an accelerating, expanding universe because redshift is caused by a decrease in the speed of light resulting from spacetime drag, not from an accelerating expansion of the universe.

Proof the Big Bang Theory Is Wrong

I am not saying that the Doppler effect does not occur with light in a similar manner as it does with sound. It certainly does. Galaxy Messier 81 (M81 or NGC3031) as shown is a good example. It is at a relatively close distance of 11.8 million light years with a super-massive black hole at the center. It has a blueshift in the light spectrum which indicates that M81 has a velocity vector toward us due to the Doppler effect. The reduction in the speed of light should have given M81 an ever-so-small redshift, but instead it has a blueshift in the light spectrum. M81 is certainly moving slowly toward us since it has a Doppler effect with a blueshift that is greater than the decrease in the speed of light. Keep in mind that the distant galaxies used in this calculation are 1000 times more distant than M81. At this vast distance, the decrease in the speed of light has 1000 times the change that it does for M81.

Galaxy Messier 81 (M81 or NGC3031)

Chapter 7

Calculation for the Reduction in the Speed of Light across Spacetime

Centuries ago scientists observed that stars emitted white light. They concluded that the speed of light was a constant. Galaxies that also emitted white light were observed for the first time through the early telescopes, confirming the assumption in the minds of astronomers that the speed of light was a constant. Astrophysicists thereafter have declared this to be the foundational law of astrophysics: "The speed of light is a constant."

Imagine yourself as a college freshman in Astronomy 101. Professor Dr. Edwin Hubble is explaining his Redshift Light Theory to the young, naive students. Suddenly, an outspoken, rambunctious male student raises his hand and blurts out at the same time, "Doctor Hubble, it seems to me that the redshift could be caused by a decrease in the speed of light as it crosses the vast universe." "No, no!" comes the immediate chastisement from Dr. Hubble. "We have known from centuries of observations and measurements that the speed of light is a constant. This is the foundational law of astrophysics." The young student submits to the professor because he knows the politics of the student-teacher relationship, but as he is leaving the room at the end of the lecture he nudges his buddy and says, "Dr. Hubble still hasn't explained how a massive galaxy could accelerate."

The brainwashing of the next generation has begun in Astronomy 101. Physicians who for 200 years rejected the theory that bacteria cause flesh infections did not set a record. Stargazers are a light-year ahead of them. It has been said that an astronomer will not discover anything new in the universe after he passes 30 years of age. Old minds have little or no flexibility. They are locked into decades of brainwashing and unable to comprehend any new object in space or any new concept. That is the nature of brainwashing. Are you brainwashed too?

We know the speed of light is not a constant because it travels much slower through transparent objects, such as glass, plastic, and air than it does through spacetime. This has been easily measured and is called the *refractive index*. The change in the speed of light through a glass or plastic prism causes the different frequencies to separate into the spectrum of colors. The same separation can be seen in a beautiful rainbow in the sky where the speed of light was changed by the falling rain as the speed and frequency decreased. In the Western United States, we see beautiful rainbows because of the low humidity. The setting sun will shine brightly in our faces as an overhead thunder cloud is drenching us.

The speed of light decreases as it travels across unimaginable billions of light-years of spacetime. The amount is so small that it was undetectable for hundreds of years until Dr. Edwin Hubble began to observe extremely distant galaxies with a new and more powerful telescope. The false assumption of the past held the speed of light constant. I will hold the wavelength of the light constant and assume that the average distant galaxy is static in spacetime. The speed of light, **C**, equals the frequency, **f**, times the wavelength, λ.

$$C = f * \lambda$$

or

$$\lambda = C / f$$

$$C_e / f_e = C_r / f_r$$

Chapter 7

Where: C_e is the emitted light speed and C_r is the received light speed.

$$C_r = C_e * f_r / f_e$$

The light was emitted from the galaxy as white light, $f_e = 0.595 \times 10^{14}$ Hz or 595 THz.

The light was received at Earth as redshift light, $f_r = 0.400 \times 10^{14}$ Hz or 400 THz.

> C_r = 186,000 miles/second * (400 / 595) or 300,000 kilometers/second.
>
> C_r = 125,000 miles/second or 201,700 kilometers/second.

Therefore, the *Rieske Spacetime Drag Coefficient (RSDC)* is the decrease in the speed of light per 1.0 billion light years.

> RSDC = (186,000 - 125,000) / 12 billion light years.
>
> RSDC = 5,080 miles per second per 1.0 billion light years,
>
> or
>
> RSDC = 8,200 kilometers per second per 1.0 billion light years.

***Rieske Spacetime Drag Coefficient* is 5,080 miles per second per 1.0 billion light years.**

Dr. Edwin Hubble should have immediately concluded that the speed of light from distant galaxies was decreasing because the redshift was equal in all directions as one would expect. He

Proof the Big Bang Theory Is Wrong

concluded instead that we are at the center of the universe which is expanding away from us in all directions—a far-fetched improbability. The odds that we are at the center of the universe are so infinitesimal that they are, in fact, zero. An analogy would be a person sailing on the ocean at some random location who looks in all directions but sees no land. He concludes he must be in the middle of the ocean because he sees no land. No! He is not at the center of the ocean simply because he cannot see land, and we are not at the center of the universe. Dr. Hubble obviously came to the wrong conclusion. Well, it is obvious to me. Now is a good time to go back to Chapter 6 and read about brainwashing again.

**M51 (NGC 5194) Hubble Remix
and Companion NGC 5195 on Left**
Credits: S. Beckwith (STScI), Hubble Heritage Team, (STScI/AURA), ESA, NASA, Additional Processing: Robert Gendler

Chapter 7

Dr. Albert Einstein's Mistake Was Claiming He Had Made a Mistake

Dr. Edwin Hubble made a big mistake by assuming that the velocity of light through spacetime remains unaffected by spacetime warp and the resulting gravitational wave. Dr. Albert Einstein made a major mistake by reluctantly accepting Dr. Hubble's faulty concept, but Dr. Einstein died never knowing or admitting that the Redshift Light Theory was false. When he developed his original *General Theory of Relativity,* Dr. Einstein's core intellect told him that the far galaxies had no means of propulsion to make them accelerate. There is nothing in the vast emptiness of space that could cause a huge galaxy with a massive black hole to accelerate in any manner. Dr. Einstein added a constant to the theory to make it conform to a non-accelerating universe based on his intuitive belief. Dr. Einstein's constant was correct, but he later said it was wrong and erased it. Erasing it was his final error.

Dr. Einstein most likely did not challenge Dr. Hubble's Redshift Light Theory because of the instant and overwhelming support from the entire community of astrophysicists. Dr. Einstein relented by erasing the constant from his theory. He confessed that the constant was a mistake, but this book now proves that his original *General Theory of Relativity* was correct. The correct reason for the redshift of light from distant galaxies as presented here would force Dr. Einstein to admit the constant was correct. His real mistake was erasing it. You can read more about this in Myth No. 8 below.

Scientists and professors may suspect that the speed of light is not a constant, but they are dead silent on this topic because they know it destroys the "holy" Big Bang Theory. A great number of nonsensical theories have been devised in attempts to redeem or

Proof the Big Bang Theory Is Wrong

prevent the crash of the false Redshift Light Theory. Since observations in the universe conflict with the Redshift Light Theory, cosmologists, astronomers, and astrophysicists must conjure up a cascading list of unproven concepts in feeble attempts to save the theory. This rabbit trail of astro-nonsense is infinite, and a new theory pops up every year like a rabbit dashing out of its hole.

Dr. Edwin Hubble Brainwashed Himself by Jumping to Hasty Conclusions—Absolute Truth Ignored

We don't know why Dr. Hubble was so hasty in giving an explanation for the redshift in the light he discovered coming from the most distant galaxies. He could have proposed a list of candidates for examination but he apparently did not. Perhaps he was influenced by the following experiences from his childhood:

- Dr. Edwin Hubble played on the railroad tracks too much as a kid.
- Dr. Edwin Hubble was falsely infatuated with the Doppler effect from railroad trains and applied the concept to light.

Astro-Nonsense Cascading from the False Redshift Light Theory

Dr. Hubble's impromptu explanation for the redshift in light has had ramifications that are monumental.

- False theory that the speed of light is a constant.
- Failure to observe the intuitively obvious fact that light, like all mass, decreases in speed while traveling through spacetime.
- False calculation that the Earth is at the center of the universe.

Chapter 7

- False calculation that the Milky Way Galaxy wasn't flung out into spacetime like everything else. It remains at the center.
- False calculation that distance galaxies are accelerating away from us.
- False calculation for the mass of the universe.
- False theory that dark matter is the "missing mass" of the universe.
- False theory that dark energy is causing the galaxies to accelerate away from us.
- False theory that negative mass and antimatter exist.
- False theory that wormholes are a possible shortcut through spacetime.
- False theory that parallel universes exist.
- Brainwashed university professors and students who watched *Star Trek: The Next Generation* too much as children or adults.

International Space Station over Earth
Credit: STS-131 Crew, Expedition 23 Crew, NASA

Proof the Big Bang Theory Is Wrong

Big Bang Flaw No. 3

The Original Ignition Dilemma

A scientific theory for the original energy source for the Big Bang is totally lacking. What event pulled the trigger to start the Big Bang? Did the universe reach a critical energy or critical mass level that started the event? The Big Bang Theory is far out. The ignition that supposedly started the Big Bang is devoid of science. There aren't any serious attempts to explain it because no explanation is possible. The original ignition did not happen because it was impossible. Somebody has been watching too many fireworks on the 4th of July. The universe has galaxies and

Chapter 7

clusters of galaxies everywhere. Some are close and some are far away. They are on every side of us. They seem to be distributed throughout the universe but in an uneven manner. This isn't logical if the universe had a common point of origin. The initial discovery that galaxies formed clusters was a great shock to astrophysicists because the Big Bang Theory predicted that galaxies would be evenly distributed at similar distances from the center. Did the astrophysicists begin to question the legitimacy of the Big Bang Theory? No! They simply ignored this observable scientific fact that has become another nail in the coffin of the Big Bang lie. When physical observations do not match the Big Bang Theory, university professors simply ignore the physical facts. These grown men with advanced degrees continue to believe that the entire universe was once concentrated in a pinpoint. This is pure nonsense.

The fireworks picture on the previous page appears to have matter distributed throughout in a fairly even manner. The picture is deceiving us because the camera film recorded the burning particles from the beginning of ignition to the end. The particles appear as streaks or curved lines rather than particles. The curved lines are caused by the gravity of the Earth. In reality, the particles in the fireworks at any instant are all very close to the same distance from the center. They form a shell or the perimeter of a sphere. All of the particles move out from the center at the same time and same distance. The universe is not like that. Somebody should have called it the *Big Squirt Theory.*

Oh, no! Are they going to tag me with the new *Big Squirt Theory* like they did the Big Bang guy?

In a 1949 radio broadcast, English astronomer, Sir Fred Hoyle (June 24, 1915 – August 20, 2001), was explaining the difference

Proof the Big Bang Theory Is Wrong

between the steady state cosmological model and an expanding universe model when he described the expanding universe as a "Big Bang." He got tagged with coining the name, which increased his stature in history and science.

Many galaxies are far from the proposed center of the Big Bang and some very close. This should not have occurred according to the Big Bang Theory. The original mass or energy would have to keep squirting out galaxies in a continuous fashion over billions of years to produce the uneven distribution we see in the universe.

> **Do you believe the universe was once a pinpoint? I hope not.**

Cosmologists and astrophysicists claim that prior to the Big Bang, the entire universe was compressed into a point smaller than one hydrogen atom at an unbelievably high temperature. Oh, really? Where did this super-compressed, super-hot universe come from? They will not even address the topic, "Where did the energy in the Big Bang come from?" Ask your professor and watch him choke. The theory that the universe was once a tiny dot that exploded in the Big Bang is insanity—mass brainwashing.

One day in the distant future, scientists and psychiatrists will ponder this question, "How did humans in the twentieth and twenty-first centuries become outrageously brainwashed on such a massive scale by the Big Bang Theory?" This brainwashing will be viewed as comparable to the brainwashing of humanity that occurred for the old *Flat Earth Theory* prior to Christopher Columbus' famous discovery of North and South America.

Chapter 7

Big Bang Flaw No. 4

The Acceleration of the Universe Conflict

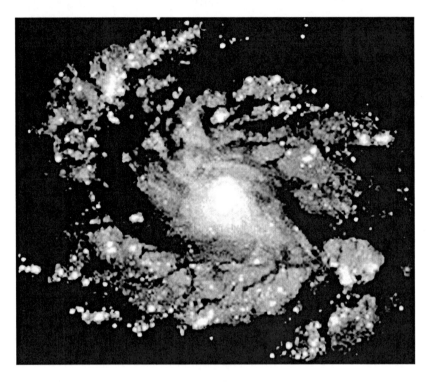

Galaxy M100

The calculated acceleration of distant galaxies and galaxy clusters is a serious dilemma for the Big Bang Theory. What is the means of the propulsion? How could a galaxy accelerate through the universe? There are no theories because the concept is silly, but it is a conclusion that one must reach if the Redshift Light Theory is true. Obviously the Redshift Light Theory is wrong. A galaxy such as M100 shown here has no possible method for increasing the speed of travel through the universe.

The surrounding mass from other galaxies, galaxy clusters, black matter, light, and unknown particles would have the effect of

Proof the Big Bang Theory Is Wrong

slowing the expansion. Masses produce a gravitational field that pulls them together. They do not push away from each other. Scientists in the past expected to see the universe expanding at a constant rate, stopped, or contracting at an accelerating rate. Expansion at an accelerating rate is impossible. The Redshift Light Theory is simply wrong.

Dr. Einstein's *General Theory of Relativity* states that the universe must be expanding. He thought the universe was static and introduced a cosmological constant as a correction factor. He later called the constant "my biggest blunder" after accepting the Redshift Light Theory that the universe is expanding. It now appears his biggest blunder was the acceptance of his first blunder. Dr. Einstein's *General Theory of Relativity* was correct with the constant.

Mt. Palomar Observatory, Palomar Mountain, CA, USA

Chapter 7

Big Bang Flaw No. 5

The Black Hole Dilemma

NASA's Illustration of a Galaxy with a Massive Black Hole
Courtesy NASA/Dana Berry, Skyworks Digital Inc.

The Black Hole Theory appears to be true. Matter at the center of a galaxy as shown here can become so concentrated that the gravitational field prevents everything from leaving. Since light cannot escape, the matter is invisible. Neither can X-rays escape. On visual observation, the black hole appears to be empty space. The black hole is not as large as the black center of this galaxy. The stars near the black hole have been drawn in. The gravitational force of a black hole causes the diameter of a star in stable orbit to be very large. Black holes have been proven to exist because of the action of stars in orbit around them and the fact that a black hole bends light that passes nearby.

According to NASA,

> "One of the most energetic explosive events known is a supernova. These occur at the end of a star's lifetime,

Proof the Big Bang Theory Is Wrong

> when its nuclear fuel is exhausted and it is no longer supported by the release of nuclear energy. If the star is particularly massive, then its core will collapse and in so doing will release a huge amount of energy. This will cause a blast wave that ejects the star's envelope into interstellar space. The result of the collapse may be, in some cases, a rapidly rotating neutron star that can be observed many years later as a radio pulsar."

A supernova has an escape velocity less than the speed of light. Therefore, light is emitted from a supernova, and matter is blasted away at a velocity greater than the escape velocity. A black hole has greater mass than a supernova. Light and matter cannot escape from a black hole because the velocity required to escape the large gravitational field is greater than the speed of light, but light and matter cannot travel greater than the speed of light. So a black hole cannot explode. All matter and light that are drawn into a black hole will remain there forever.

Black Holes cannot explode.
The Big Bang could not have exploded either.

The origin of the universe just prior to the proposed Big Bang would have been trillions of trillions of times more massive than any suspected black hole. We could think of this center as having infinite mass and an infinite gravitational field. Nothing could have escaped. The matter could never have expanded. No explosion of any type could have produced the power to blow the matter apart. Energy would have been contained by the massive gravitational field in the same way that the energy of a black hole is contained. Electrons could not have circled the nucleus of an atom. All electron motion would have been pulled to a stop by the massive gravitational force. The space between electrons,

Chapter 7

protons and neutrons would have become zero. Atoms as we understand them would not have existed.

Black holes also prove that an enormous amount of mass could not exist at the size of a pinhead in any form at the instant of the Big Bang as is falsely claimed by Big Bang proponents and college professors. Black holes have a large diameter that is proportional to the amount of mass. The minimum size of a black hole is called the *Rieske Minimum Black Hole Mass.* This size is reached when the gravitational force prevents light from escaping. The minimum diameter for the smallest possible black hole is certainly massive—not the size of a pinhead. A pinhead-size black hole is science-fiction nonsense.

Measurements of known black holes have shown that the diameter of the black hole varies in proportion to the mass. These measurements are taken by observing the visible mass (stars) which are in orbit around the black hole. Matter has a maximum density much less than that claimed by the Big Bang proponents. All matter in the universe could not have existed in a space the size of a pinhead as most Big Bang scientists claim. That is simply nonsense without any scientific basis whatsoever. The *Rieske Minimum Black Hole Mass* can be easily observed throughout the universe, and it proves the Big Bang Theory is false.

Proof the Big Bang Theory Is Wrong

Big Bang Flaw No. 6

The Big Bang Residue Went Missing

A supernova is a massive star that loses much of its mass when it collapses and explodes. The residue remaining after the explosion becomes a rapidly-rotating neutron star, or radio pulsar.

Something should be left as residue from the Big Bang if the universe had been created in this way, but the residue cannot be found. The residue should be relatively easy to find if it existed because the Earth is said to be near the center of the universe, not far away in an edge galaxy.

> **Astrophysicists made a big mistake by treating the Big Bang Theory as Law. Now they can't correct the blunder.**

Comparing a car bomb to the Big Bang Theory is another good analogy. The car filled with explosives is at the center of the blast. Objects close to the car are blown far away, but the residue of the car remains exactly where the blast occurred. The remaining residual mass from the Big Bang should have collapsed to form a massive object. Other matter near the center should have been seen as a spiral collapsing into the center. Nothing can be found. It has either gone missing or it never existed.

The lack of physical evidence at the scene proves the Big Bang did not occur. The great television homicide detective, Lieutenant Colombo with the Los Angeles Police Department, would have concluded that the Big Bang never existed. There is no residual matter to be found that should have been left behind.

Chapter 7

There is no massive black hole or any other object to identify the Big Bang origin. The Big Bang Theory is simply a myth that cannot stand up to a close examination of the physical evidence.

Low-level background radiation has been measured by astrophysicists, but it does not match the intensity or frequency expected from the Big Bang. The background radiation is simply coming from the billions of galaxies throughout the universe.

Cosmologists who support the Big Bang Theory propose that background or microwave radiation would have remained from the initial explosion. They were elated when radiation similar to that expected was accidentally discovered in 1965. They thought this certainly confirmed the Big Bang Theory, but it did not. The theory is still in a shambles because the background radiation has gone missing.

The background radiation discovered is certainly not from the Big Bang. It is simply radiation produced by the stars, galaxies, and other matter in the universe. The radiation simply does not fit the requirements for the Big Bang.

- The background radiation is omni directional. It comes from every direction possible, not from the suspected Big Bang point of origin.
- The background radiation is too weak. The Big Bang should have produced a much stronger radiation measurement. Instead, it fits the radiation expected from galaxies.
- The background radiation is too smooth. It is very even and does not fit the Big Bang Theory.
- The background radiation temperature spectrum is too low.
- The background radiation has the wrong spectrum. It does not match the black body spectrum as expected.

Proof the Big Bang Theory Is Wrong

Rather than supporting and confirming the Big Bang Theory, the background radiation adds more doubt. The background radiation matches what one would expect to come from the galaxies of the universe. The background radiation expected from the Big Bang Theory does not exist because the event never happened.

Wait a minute! Dr. Edwin Hubble told us that the Earth is at the center of the universe. If this is true, we should be at the center of the residue radiation. If the Big Bang Theory were true, we would be fried by leftover energy.

Chapter 7

Big Bang Flaw No. 7

The Impossible Computer Calculation

Scientists and computer geniuses cannot model the Big Bang Theory because the computer would simply say, "Cannot compute." Einstein's *General Theory of Relativity* and other laws of science cannot be used in a computer program to model the Big Bang because the computer would simply say, "The Big Bang Theory is a myth." Computer programs that claim to simulate the Big Bang are frauds.

Computer programs cannot be written to simulate the Big Bang because the scientific laws to govern the equations are nonexistent. The programs must violate proven physical laws of the universe and Dr. Albert Einstein's *General Theory of Relativity*. Astrophysicists have brainwashed themselves into a corner.

Proof the Big Bang Theory Is Wrong

Big Bang Flaw No. 8

The Colliding Galaxies Dilemma

Spiral Galaxies NGC2207 and IC2163

We can see from this picture that the two galaxies appear to be colliding. Many galaxy pairs are seen this way. They appear to be exchanging matter. We should accept the obvious.

> **The Hubble Space Telescope (HST) decimates the Big Bang Theory.**

Some galaxies appear to be nearing a collision. Other galaxies have been interpreted as nearing a collision, but upon closer examination we can see they are not. The difference is in the detail of the stars. Let's say you use a telephoto lens to take a picture of a train 10 miles (16.7 km) away traveling toward you. A man walking on the railroad tracks one mile (1.7 km) ahead of the train appears in the picture to be in imminent danger. The same deception can be seen with galaxies. One is behind the other.

Chapter 7

Cosmologists tell us that some galaxies are traveling toward each other and are going to collide or have a near miss. This simply could not be true if they came from the same point of origin as the Big Bang Theory states. Galaxies cannot change direction. The scientific law of the conservation of momentum proves galaxies cannot change direction. According to the Big Bang Theory, galaxies should all be expanding away from each other, not moving, or all moving toward each other. Cosmologists say some are moving away from each other and some are moving toward each other. Somebody here is seriously wrong. The two colliding galaxies prove the Big Bang Theory is wrong.

$$F = M * A$$

Sir Isaac Newton (1642 - 1727) was one of the giants of science. Newton's Second Law of Motion has never failed in theory or in observation. Newton's second law can be restated as force **(F)** equals the mass **(M)** multiplied by the acceleration **(A)** rate. This formula is very simple, but it is the foundation for calculations that engineers and scientists use every day.

> **A galaxy can't make a right-hand turn.**
> **A galaxy can't make a left-hand turn either.**
> **Colliding galaxies destroy the Big Bang Theory.**

This equation means that the force applied to any object will cause it to accelerate. A planetary space probe will continue to travel in a straight line unless a force is applied to change the speed or change the direction. Rocket side thrusters can be fired to make a mid-course correction. Rocket thrusters built into the front of the space probe can be fired in the reverse direction to slow the space probe for orbit around a planet.

Proof the Big Bang Theory Is Wrong

Kepler's laws of planetary motion defined the gravitational relationship between two objects. Gravity is something we all understand when we fall off a ladder. Each particle of matter on Earth and in the universe attracts every other particle with a force that is directly proportional to the product of their masses and inversely proportional to the square of the distance between them. The formula below is the force developed by this attraction. The force accelerates our fall from the ladder until we hit the ground. You will not one day go floating off into space in some other direction.

$$F = (G * M_1 * M_2) / D^2$$

The force (**F**), newton, equals the product of the gravitational constant (**G**) times the mass of the first object (**M₁**), kilogram, times the mass of the second object (**M₂**), kilogram, divided by the square of the distance (**D**), meter, between the centers of gravity of the two masses. The gravitational constant (**G**) in Newton's formula has the value:

$$G = 6.672 \times 10^{-11} \text{ N m}^2 \text{ kg}^{-2}$$

The weight you see when you step on a scale is the force between you, the Earth, and everything else in the universe. The space probe in our example above is attracted by everything else in the universe as well. The probe enters an orbit around another planet when the gravitational force of the planet equals the centrifugal force required to travel in a circle. These two laws can be proved by spinning a mass on the end of a wire or by observing any galaxy in the universe. Stars are held in a precise orbit within a galaxy because these two laws are exact and true in every galaxy of the universe.

Chapter 7

In his *General Theory of Relativity*, Dr. Einstein proposed that both distance and time are flexible but form a single entity he called *spacetime*. He proposed that the time variable in spacetime causes a very small deviation from Newton's formula that ignored time. Many modern scientists and science writers have the audacity to claim that Newton's formula proposed in 1666 is wrong. Newton's formula was one of the most brilliant scientific achievements known to mankind. His formula is still used to this day in placing space probes into proper orbit around distant planets. A formula for Dr. Einstein's gravity calculations cannot be found because it is simply too complex to be practical, but Dr. Einstein's *General Theory of Relativity* does provide a more accurate scientific understanding for gravity based on spacetime. Why some science writers hate and harshly criticize Newton is beyond comprehension. Perhaps it is because Newton believed the universe was created by God. The correction of Newton's formula involves changing the exponent of the radius from 2.0 to 2.00000016. This is less than two parts in ten million. Modern scientists have validated this change based on observations in which:

$$G = 6.672 \times 10^{-11} \text{ N m}^2 \text{ kg}^{-2.00000016}$$

Dr. Einstein proved that gravity changes the direction of light, but he failed to theorize that gravity can increase or decrease the speed of light. Because of this misunderstanding, he accepted Hubble's Redshift Light Theory and deleted his constant from his gravity calculations. He believed that his original *General Theory of Relativity* was wrong. This book will prove that Dr. Einstein's original theory was right. Dr. Einstein half-heartedly accepted the Big Bang Theory. Dr. Einstein's original *General Theory of Relativity* showed the Redshift Light Theory to be theoretically wrong.

Proof the Big Bang Theory Is Wrong

The difference between Sir Isaac Newton's gravity formula and Dr. Albert Einstein's formula gives us the answer for the redshift of light from distant galaxies. Newton's formula lacks a factor for the resistance of mass traveling through spacetime. Dr. Einstein's formula tells us that the velocity of mass decreases because of the resistance of mass traveling through spacetime. Light has mass. Light must slow at it travels through spacetime. The speed of light is not constant as is taught in universities around the world. The little factor with six zeros plus 16 literally vaporizes the Big Bang Theory.

$$G = 6.672 \times 10^{-11} \text{ N m}^2 \text{ kg}^{-2.00000016}$$

This is a very important scientific discovery. I call this new discovery the *Rieske Spacetime Drag Coefficient for the Speed of Light*. I used the original issue of Dr. Albert Einstein's *General Theory of Relativity* to validate this new discovery for the redshift in the spectrum of light coming from distant galaxies. *The Rieske Spacetime Drag Coefficient for the Speed of Light* proves beyond any possible doubt that Dr. Edwin Hubble's Redshift Light Theory is wrong, and in doing so destroys the Big Bang Theory as well. This new Law of Science proves that the speed of light slows as it travels through the universe due to the resistance that spacetime imposes on mass. The decay in the velocity of mass traveling through spacetime has already been confirmed by observing the decay in the orbital time of binary pulsar PSR 1913+16. The distant galaxies are not accelerating away from Earth as astrophysicists and cosmologists have claimed. The speed of light is not a constant as is stated in science books and taught by university professors. The universe is not expanding. Black Energy does not exist. The Big Bang Theory is a dead corpse.

Chapter 7

Big Bang Flaw No. 9

The Expanding Gas Cloud Dilemma

Gaseous Pillars M16
Credit: James Long & the ESA/ESO/
NASA Photoshop FITS Liberator

Big Bang enthusiasts tell us that the original explosion sent out clouds of hydrogen and helium gases which compressed and coalesced into galaxies and stars. There is a big problem with this

Proof the Big Bang Theory Is Wrong

idea. Clouds of gases don't compress and coalesce into a more dense mass. The scientific truth is the opposite. Gases expand and become less dense. The gravitational force between gas molecules is smaller than the molecular diffusion force as determined by Fick's Law of Diffusion. Clouds of gases in the universe similar to those of Gaseous Pillars M16 pictured on the previous page can be seen with the Hubble telescope. These clouds are not coalescing or collapsing into a more dense mass. Stars cannot be formed this way. There is not enough gas in a given space to make a star. The gas would simply drift away and dissipate. Obviously, the gases in this picture of the universe are not coalescing into a star or galaxy. They maintain their odd shapes because of the very low gravitational forces. The gases would form a spherical shape if they were compressing and coalescing into a more dense mass. This is just simple science based on proven scientific laws. The picture proves the Big Bang Theory is wrong.

Our smaller planets prove that gas clouds could not form stars. The planet Mercury does not have an atmosphere because gases simply float away into space. The planet does not have enough gravity to keep any gas molecules it temporarily captures. Our Moon is another example. A cloud of gas released on the surface of the Moon would simply float off into space.

A gas cloud could not form a galaxy, star, or planet. The outer fringes of the cloud would dissipate, and the cloud would get smaller and smaller. The only possibility is that the galaxy, star, or planet had dense matter from the beginning. The gas cloud theory is a bunch of hot air.

The Big Bang Theory is a dead corpse, but your university professor will insist that you embrace it.

Chapter 7

Many of the observations that disprove the Hubble Redshift Light Theory have been made by the very space telescope that bears his name. This is truly ironic.

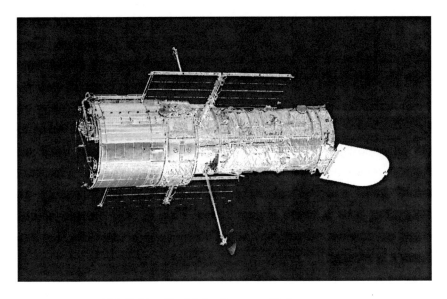

NASA's Hubble Space Telescope

Hubble Essentials

http://hubblesite.org/the_telescope/hubble_essentials/

Hubble Nuts and Bolts

http://hubblesite.org/the_telescope/nuts_.and._bolts/

Proof the Big Bang Theory Is Wrong

Big Bang Flaw No. 10

Expanding Space Addendum Falters and Fails

Although its validity is totally lacking, cosmologists push the Big Bang Theory into school curricula as if it were truth. A new addendum has been proposed in an attempt to keep the Big Bang Theory from utter collapse. The expanding space addendum is basically an admission that the Big Bang Theory has many faults. The expanding space concept is an attempt to resolve them.

Chapter 7

The expanding space idea suggests that the Big Bang was not a gigantic explosion but rather an expansion of space. The proponents claim the speed of light is not a limitation in the expanding space concept because expanding space has no speed limitation. The speed of light is only a limitation within the boundaries of space. This proposal suggests that space has edges, and if you go too far you will fall off. This concept is reminiscent of the old flat Earth idea commonly accepted several centuries ago. Sailors refrained from venturing too far out to sea for fear of falling off the edge of the flat Earth. Things never change.

The expanding space concept has many flaws. Space is not a physical body with limits. Space is not something that can move or have dimensions. Space is not expanding. Space is infinite. Space and time extend forever as spacetime.

Another major flaw in the expanding space concept can be seen in distant galaxies. If space were expanding, we should see the size of distant galaxies expanding as well. Space exists between the stars that make up a galaxy. The stars within the galaxy should be farther apart in distant galaxies than in close ones but they aren't. The expanding space proponents simply sidestep this problem by saying that space expands all around a galaxy but not within the galaxy. No! This logic does not compute. The expanding space theory has no scientific or observational support. Space cannot be seen, felt, measured, or observed because space is spacetime. There are no forces available to stretch spacetime.

Proof the Big Bang Theory Is Wrong

The Cross Galaxy

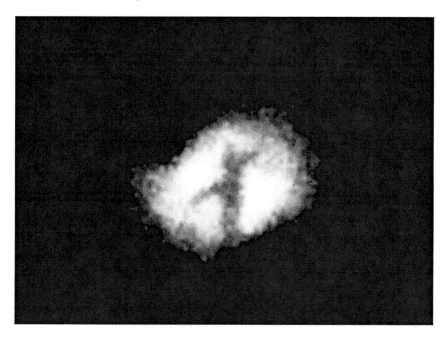

Center of M51 Whirlpool Galaxy

This picture of the center of the M51 Whirlpool Galaxy, sometimes called the *Cross Galaxy*, was taken by the Hubble Telescope. The M51 Whirlpool Galaxy should certainly convince you that Jesus Christ was God manifest in the flesh and that Christ created the universe. Jesus Christ created the M51 Whirlpool Galaxy with an exact image of His cross in the middle. The Bible makes this very plain.

> **Colossians 1:15** "He is the image of the invisible God, the firstborn over all creation. 16 For by Him all things were created that are in heaven and that are on earth, visible and invisible, whether thrones or dominions or principalities or powers. All things were created through Him and for Him. 17 And He is before all things, and in Him all things consist."

Chapter 7

In the beginning God created the Heavens and the Earth.

The stretching of the heavens as described in the Bible may simply mean that God spread the galaxies across the universe over a vast expanse that we see with modern telescopes.

Genesis 1:1 "In the beginning God created the Heavens and the Earth."

Jeremiah 10:12 "He has made the earth by His power, He has established the world by His wisdom, And has stretched out the heavens at His discretion."

Proof the Big Bang Theory Is Wrong

Big Bang Flaw No. 11

Time, Distance, and Age Are All Messed Up

The Big Bang Theory demands that older galaxies be toward the outer edge of the universe; however, this is not the case. Galaxies very far away should be old, though some appear to be young. Galaxies close to us should be young since we are said to be near the center of the Big Bang, but some appear to be old. The Big Bang Theory is simply a mess.

Modern super computers can estimate the distance of nearby galaxies based on their size—as done with a hunter's rifle telescope to judge the distance of a deer. This is called the *minute of angle (moa)* method. Galaxies of similar appearance are a similar size as are people and deer. The size is inversely proportional to the distance from the Earth. In other words, a galaxy that fills the width of the telescope is half as far away as another galaxy that only fills the telescope to half the width.

Other redshift-independent methods have been used to calculate the distance to nearby objects of the universe. The NASA Astrophysical Data System (ADS) published on January 1, 2009, lists distances to approximately 10,000 galaxies using 30,000 redshift-independent methods. The methods that calculated distance within 10% accuracy included Cepheids, Type-Ia Supernovae, and Tip of the Red Giant Branch using standard candles methods. Other standard rulers include Masers, Globular Cluster radii, and others. The Tully-Fisher, Fundamental Plane, and other less accurate space measuring methods within 20% accuracy were listed.

Chapter 7

Micro or Mini Black Holes and Dark Energy

Astrophysicists have invented what they call *Micro Black Holes*. This is the craziest theory of the century . Micro black holes are said to be tiny black holes. One astrophysicist on TV claimed, "I could have a mini black hole on the end of my nose at this moment." The theory of a mini black hole is so stupid that I will make no attempt to define it. The concept of a mini black hole destroys the logic, facts, and understanding of a true black hole—a mass of matter so large that the gravitational force created prevents light from escaping.

> **Proverbs 12:1** Whoever loves instruction loves knowledge, but he who hates reproof is stupid.

Let me pause to explain Bible verse Proverbs 12:1. It means that if anyone hates this book, he is well . . . . . not so bright.

Forty years ago, I heard astrophysicists claim that planets may not exist in orbit around other stars—only our Sun. I thought the astrophysicists had gone mad. I replied, "Certainly there is not even one star in the galaxy without planets and massive amounts of matter in orbit around it." Our Sun has numerous planets, and many of the planets have numerous moons. The Sun has a belt of asteroids and unnumbered comets, asteroids, gas, dust, and everything else imaginable. Planets must be in orbit around every star in the Milky Way Galaxy.

Now the astrophysicists have gone wacko in the opposite direction. They were very slow to accept the obvious prediction that planets were in orbit around other stars. They would not accept the obvious until they could see visible proof with their eyes. Now they invent absolutely nutty ideas like micro black holes in order to explain the obvious faults of the erroneous Big Bang Theory.

Proof the Big Bang Theory Is Wrong

Dark Energy is another nonsensical idea presented by astrophysicists in an attempt to keep the Big Bang Theory from collapsing. The Big Bang Theory simply does not match the observable facts that are discovered daily. Dark energy is akin to Bigfoot—the imaginary ape-like humanoids. Gullible and brainwashed people believe in them, but they can't find even one. Astrophysicists claim dark energy is causing the universe to expand at an accelerating rate even though they can't find it. They arrive at this unproven idea because the Redshift Light Theory is also wrong. One error leads to ten more.

Multiverse, Meta-Universe, Multiple Universes, Parallel Universes, and Space Wormholes

Astrophysicists are claiming that multiple universes exist. These multiple universes may be identical twins to our real universe, where copies of each of us live an identical life. A variation of this theory is a parallel universe that exists within our real universe. Other theories suggest that each of the multiple universes is different.

Astrophysicists have proposed the theory of a space wormhole, which they claim is a shortcut path to a far point across the universe or to a parallel universe. The space wormhole bypasses the barrier that matter cannot travel faster than the speed of light. The "wormhole" theory is obviously wrong. The common description of a wormhole through spacetime goes something like this:

> "A two-dimensional visualization of spacetime is shown folded like a piece of paper until the two distant sides of the paper are near each other. A hole exists in one end of the sheet and connects to a similar hole in the other end

Chapter 7

of the sheet. This hole allows travel from one side of spacetime to the other side in a short distance."

The analogy does not compute. Spacetime is not two-dimensional. It is four-dimensional, comprised of three dimensions of physical space plus time. The four dimensions of spacetime cannot be folded like a sheet of paper. Give it up! Wormholes don't even make good science fiction movies much less subjects in college curricula.

2-D Illustration of the Wormhole Theory

Do not believe these theories. They are nonsense. These concepts originated in science fiction movies and TV programs like *Star Trek: The Next Generation*. Unfortunately, many astrophysicists watched these programs as children, and they have become brainwashed to the extent that they can no longer distinguish reality from fiction.

Proof the Big Bang Theory Is Wrong

Astrophysicists believe multiple universes and space wormholes are realities, but they reject observations that the speed of light decreases as it travels through spacetime. They reject the obvious slowing of the speed of light because accepting it would instantly destroy the Big Bang Theory. Astrophysicists have not gone awry by accident. They intentionally brainwash themselves and others.

Astrophysics and cosmology are no longer sciences. Like biology and human physiology, astrophysics and cosmology stumble over the obvious and proclaim unproven theories as facts. Science fiction writers have now moved into the universities where they fabricate and invent extreme nonsense at the expense of the students and tax payers. Modern biology is based on the false theory of the evolution of the species. Human physiology is based on the false theory that the low-fat diet is best. Modern astrophysics is based on the false theories that an expanding universe causes a redshift in the light; the universe was created by a Big Bang from a pinpoint of energy; space wormholes are shortcuts through the universe; parallel identical universes exist where we are all multiple copies; and other nonsense.

Herbert George (H. G.) Wells and Jules Gabriel Verne wrote science fiction that became very popular because they had a good understanding of true science. The fiction was close enough to reality to make the reader believe it could be possible. Today astronomers and astrophysicists venture beyond science fiction. The wildest concepts, such as wormholes, antimatter, and parallel universes are readily accepted without even a thought of reality.

Chapter 7

Absolute Truth about the Age of the Earth Controversy

Based on a literal interpretation of the Bible's six "days" of creation, many Christians believe the Earth is about 6,000 years old. This doctrine is false because the word *days* translated by Hebrew scholars into Greek as *yowm* {yome} in Genesis 1:5, can mean yesterday, today, tomorrow, time period, or age. Scientific geological features give evidence that the Earth has existed in the present form for a long time. The Greenland Ice Cap is more than 100,000 years old; ancient sea coral deposits reveal the Earth's rotation around the Sun in ages past was 400 days per year; the Moon has stopped rotating in relation to the Earth and is always facing toward us; and the meteor impact craters on the Earth have disappeared over the ages. Geological evidence shows the Earth has experienced many ages of heating and cooling in which the Arctic Ice Cap extended as far south as New York City.

The Young Earth Creationists are simply wrong. The young Earth proponents claim that God made the geological features to **appear** old. This is blasphemy toward God since it accuses Him of being a deceiver and a liar. In the beginning, God created the universe and the Earth as stated in Bible verse Genesis 1:1. This was the Earth with dinosaurs that He obliterated in Genesis 1:2. In Genesis 1:3, God recreated the Earth as we see it today. Each of these periods, ages, or "days" could have been millions of years long. This interpretation of the series of events for the creation has come to be known by Christians as the *Gap Theory*. A detailed description of events can be read in *The Bible, Genesis & Geology* by Gaines Johnson.

I know God created mankind and the newer animals more recently, within the last 10,000 years or so because the geological record is nonexistent for structures, foundations, tools, or human remains as recent as only 20,000 years ago.

Proof the Big Bang Theory Is Wrong

My Predictions for the Big Bang Theory

Before I could write my predictions for the Big Bang Theory, I first had to understand the ground rules behind its origin and continued support. Big Bang is false god No. 2. Big Bang created the entire universe in a flash of light that blew galaxies billions of light years across the universe at speeds multiple times the speed of light. Big Bang is the fastest false god to have ever been invented. The Big Bang Theory is a religion.

Evolution is false god No. 1. Evolution created all life forms on Earth but was painfully slow to accomplish this. Millions of years were required for evolution to make an eyelid on a frog. The Theory of Evolution is a religion.

The scientific élite had to invent the two major false gods in order to more conveniently deny God Almighty, Creator of all things. People defend their religions with vigor and endless energy. This is the key to making the following predictions:

- My theory above, in which strong scientific evidence is presented to show that the speed of light decays as it flows across billions of light years of the universe, will simply be ignored. In fact, major efforts will be used to ignore my discovery. It will be ignored to the point of no response.
- The scientific élite have invested their lives and careers in the Big Bang. They have been paid to complete study after study, present lecture after lecture, and write book after book describing the Big Bang Theory. They will not throw all of these away quickly.
- Perhaps as many as one million web pages and thousands of books will be obsolete if astrophysicists uniformly

Chapter 7

- agree that the speed of light decreases as it travels billions of light-years across spacetime.
- University professors and teachers will ridicule and insult any student who attempts to raise a question of doubt that the speed of light is a constant. The professors will slam the discussion to a halt, silence the questioning student, and immediately change the subject. No debate will be considered. Students have already written emails to me confirming the attacks.
- The majority of the other students who are over their heads in brainwashing will join in the attack by sneering, faking boisterous laughter, hissing, and blurting out insults against the lonely student who has a mind willing to seek the truth. This will be major "put-down" time on campus.
- The military complexes in Russia and China will secretly develop programs to determine if the speed of light from distant galaxies has decayed. They will find it to be true and attempt to use this knowledge for some strategic advantage over the United States, but they will find little application for the information at the present time. The Russians and Chinese will keep quiet about the subject in hopes of finding a strategic advantage in the future. The United States military studied the UFO myth many years ago, but they will ignore the truth that the speed of light is not constant.
- In 100 years (or maybe 200), the scientific élite will make a major announcement, "The speed of light is not constant."

Absolute Truth Exposed

Volume 1

Chapter 8

Bible Interpretation Rightly Divided
A Bible Summary

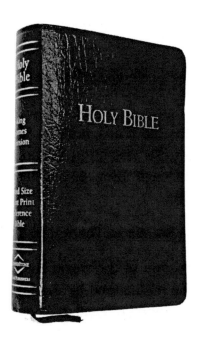

Who Is God?

Genesis 1:1 In the beginning God created the heavens and the earth.

Genesis 1:11 Then God said, "Let the earth bring forth grass, the herb that yields seed, and the fruit tree that yields fruit according to its kind, whose seed is in itself, on the earth"; and it was so.

Chapter 8

Genesis 1:24 God said, "Let the earth bring forth the living creatures according to its kind: cattle and creeping thing and beast of the earth, each according to its kind"; and it was so.

Genesis 1:26 Then God said, "Let Us make man in Our image, according to Our likeness; and let them have dominion over the fish of the sea, over the birds of the air, and over the cattle, over all the earth and over every creeping thing that creeps on the earth."

Genesis 1:27-28 So God created man in His own image; in the image of God He created him; male and female He created them. Then God blessed them, and God said to them, "Be fruitful and multiply; fill the earth and subdue it; have dominion over the fish of the sea, over the birds of the air, and over every living thing that moves on the earth."

Exodus 3:14 And God said to Moses, "I AM WHO I AM"; and He said, "Thus you shall say to the children of Israel, 'I AM has sent me to you.'"

Psalms 145:18 The LORD is near to all who call upon Him, to all who call upon Him in truth.

John 1:1-3 In the beginning was the Word, and the Word was with God, and the Word was God. He was in the beginning with God. All things were made through Him, and without Him nothing was made that was made.

Psalms 14:1 The fool has said in his heart, "There is no God."

Romans 1:20 For since the creation of the world His invisible attributes are clearly seen, being understood by the things that are made, even His eternal power and Godhead, so that they are without excuse.

A Bible Summary

Isaiah 40:28 Have you not known? Have you not heard? The everlasting God, the LORD, the Creator of the ends of the earth, neither faints nor is weary. There is no searching of His understanding.

Psalms 96:11-13 Let the heavens rejoice, and let the earth be glad; let the sea roar, and all its fullness; let the field be joyful, and all that is in it. Then all the trees of the woods will rejoice before the LORD. For He is coming, for He is coming to judge the earth. He shall judge the world with righteousness, and the peoples with His truth.

Psalms 47:2-3 For the LORD Most High is awesome; He is a great King over all the earth. He will subdue the peoples under us, and the nations under our feet.

Psalms 47:8-9 God reigns over the nations; God sits on His holy throne. The princes of the people have gathered together, the people of the God of Abraham. For the shields of the earth belong to God; He is greatly exalted.

Psalms 23:1-4 The LORD is my shepherd, I shall not want. He makes me to lie down in green pastures; He leads me beside the still waters. He restores my soul; He leads me in the paths of righteousness for His name's sake. Yea, though I walk through the valley of the shadow of death, I will fear no evil; for You are with me; Your rod and Your staff, they comfort me.

Psalms 5:4-6 For You are not a God who takes pleasure in wickedness, nor shall evil dwell with You. The boastful shall not stand in Your sight; You hate all workers of iniquity. You shall destroy those who speak falsehood; the LORD abhors the bloodthirsty and deceitful man.

Chapter 8

Psalms 11:5 The LORD tests the righteous, but the wicked and the one who loves violence His soul hates.

Psalms 34:16 The face of the LORD is against those who do evil.

Proverbs 6:16 These six things the LORD hates, yes, seven are an abomination to Him: a proud look, a lying tongue, hands that shed innocent blood, a heart that devises wicked plans, feet that are swift in running to evil, a false witness who speaks lies, and one who sows discord among brethren.

Commentary:

God is the Creator of all things: the heavens (stars, planets, etc.), Earth, all animals, and mankind. God is Triune, composed of three persons whom the Bible calls the *Godhead*. The three are God the Father, God the Son, and God the Holy Spirit. The Bible often refers to God in the plural tense.

> **Genesis 1:26** Then God said, "Let Us make man in Our image, according to Our likeness; let them have dominion over the fish of the sea, over the birds of the air, and over the cattle, over all the earth and over every creeping thing that creeps on the earth."

God placed man in charge of the Earth and over every living thing that lives on the Earth. God was in the beginning. He is eternal. He is holy, righteous, faithful, everlasting, infinite, all-powerful, ever-present, all-knowing, all-wise, absolutely just, almighty, long-suffering, loving, and merciful. He is our Creator, Judge, King, Defender, Preserver, and Shepherd. God is Love. He is the Most High God. God is awesome. But God hates those who continually reject Him to do evil until their day of judgment at God's White Throne. They will then be thrown into the eternal Lake of Fire.

A Bible Summary

Romans 1:24 Therefore God also gave them up to uncleanness, in the lusts of their hearts, to dishonor their bodies among themselves, 25 who exchanged the truth of God for the lie, and worshiped and served the creature rather than the Creator, who is blessed forever. Amen. 26 For this reason God gave them up to vile passions. For even their women exchanged the natural use for what is against nature. 27 Likewise also the men, leaving the natural use of the woman, burned in their lust for one another, men with men committing what is shameful, and receiving in themselves the penalty of their error which was due. 28 And even as they did not like to retain God in [their] knowledge, God gave them over to a debased mind, to do those things which are not fitting; 29 being filled with all unrighteousness, sexual immorality, wickedness, covetousness, maliciousness; full of envy, murder, strife, deceit, evil-mindedness; [they] [are] whisperers, 30 backbiters, haters of God, violent, proud, boasters, inventors of evil things, disobedient to parents, 31 undiscerning, untrustworthy, unloving, unforgiving, unmerciful; 32 who, knowing the righteous judgment of God, that those who practice such things are deserving of death, not only do the same but also approve of those who practice them.

Revelation 20: 11 Then I saw a great white throne and Him who sat on it, from whose face the earth and the heaven fled away. And there was found no place for them. 12 And I saw the dead, small and great, standing before God, and books were opened. And another book was opened, which is [the] [Book] of Life. And the dead were judged according to their works, by the things which were written in the books. 13 The sea gave up the dead who

Chapter 8

were in it, and Death and Hades delivered up the dead who were in them. And they were judged, each one according to his works. 14 Then Death and Hades were cast into the lake of fire. This is the second death. 15 And anyone not found written in the Book of Life was cast into the lake of fire.

A Bible Summary

Who Is Mankind?

Isaiah 53:6 All we like sheep have gone astray; we have turned, every one, to his own way.

Isaiah 64:6 But we are all like an unclean thing, and all our righteousnesses are like filthy rags.

Romans 7:24-25 O wretched man that I am! Who will deliver me from this body of death? I thank God - through Jesus Christ our Lord!

Romans 10:3 For they being ignorant of God's righteousness, and seeking to establish their own righteousness, have not submitted to the righteousness of God.

Ecclesiastes 7:20 For there is not a just man on earth who does good and does not sin.

Proverbs 16:25 There is a way which seems right to a man, but its end is the way of death.

Proverbs 28:5 Evil men do not understand justice, but those who seek the Lord understand all.

Proverbs 12:1 Whoever loves instruction loves knowledge, but he who hates reproof is stupid.

Proverbs 12:4 An excellent wife is the crown of her husband, but she who causes shame is like rottenness in his bones.

Proverbs 14:2 He who walks in his uprightness fears the LORD, but he who is perverse in his ways despises Him.

Proverbs 15:9 The way of the wicked is an abomination to the LORD, but He loves him who follows righteousness.

Chapter 8

Proverbs 31:30 Charm is deceitful and beauty is vain, but a woman who fears the LORD, she shall be praised.

James 4:4 Adulterers and adulteresses! Do you not know that friendship with the world is enmity with God? Whoever therefore wants to be a friend of the world makes himself an enemy of God. (God considers friendship with worldly people to be adultery. He called Israel a harlot for idol worship. Ezek 23:30 and Hosea 4:12.)

Proverbs 5:3-5 For the lips of an immoral woman drip honey, and her mouth is smoother than oil; but in the end she is bitter as wormwood, sharp as a two-edged sword. Her feet go down to death, her steps lay hold of hell.

Proverbs 30:20 This is the way of an adulterous woman; she eats and wipes her mouth, and says, "I have done no wickedness."

Jeremiah 17:9-10 "The heart is deceitful above all things, and desperately wicked; who can know it? I, the LORD, search the heart, I test the mind."

Romans 1:24-32 Therefore God also gave them up to uncleanness, in the lusts of their hearts, to dishonor their bodies among themselves, who exchanged the truth of God for the lie, and worshiped and served the creature (earth and animals) rather than the Creator, who is blessed forever. Amen. For this reason God gave them up to vile passions. For even their women exchanged the natural use for what is against nature (lesbians). Likewise also the men, leaving the natural use of the woman, burned in their lust for one another, men with men (homos) committing what is shameful, and receiving in themselves the penalty of their error which was due (AIDS). And even as they did not like to retain God in their knowledge, God gave them over

A Bible Summary

to a debased mind, to do those things which are not fitting; being filled with all unrighteousness, sexual immorality, wickedness, covetousness, maliciousness; full of envy, murder, strife, deceit, evil-mindedness; they are whisperers, backbiters, haters of God, violent, proud, boasters, inventors of evil things, disobedient of parents, undiscerning, untrustworthy, unloving, unforgiving, unmerciful; who, knowing the righteous judgment of God, that those who practice such things are worthy of death, not only do the same but also approve of those who practice them.

Commentary:

Everyone is born with a sinful nature, and since the fall of Adam and Eve, all people have been born without God's Holy Spirit within them. When Adam and Eve disobeyed God, all of mankind became separated from God. Jesus Christ reversed this separation by His death upon the cross. This is called the reconciliation of mankind back to God. Jesus reversed the separation caused by Adam, but this does not mean everyone will go to Heaven. Reconciliation is not Salvation. Everyone's sins have been forgiven, but not everyone will be saved from eternal punishment. If you believe in Jesus and believe in your heart that God raised Him from the dead, you will be saved.

Many sincere people have tried to become right in God's sight through self-improvement, good deeds, ethics, philosophy, religion, tradition, and obedience to the Ten Commandments. Seeking to establish one's own righteousness is not acceptable to God according to Romans 10:3. This self-effort is contrary to God's plan and has not brought about the love and abundance of life we all desire. We cannot work our way into God's favor by our own efforts.

Chapter 8

What Is the Law?

The Ten Commandments
(The Mosaic Law)
Exodus 20:1-17

I
I am the LORD your God. You shall have no other gods before Me.

II
You shall worship Me only; for I, the LORD your God, am a jealous God.

III
You shall not take the name of the LORD your God in vain.

IV
Remember the Sabbath day, to keep it holy.

V
Honor your father and your mother.

VI
You shall not murder.

VII
You shall not commit adultery.

VIII
You shall not steal.

IX
You shall not bear false witness against your neighbor.

X
You shall not covet your neighbor's wife, nor anything that is your neighbor's.

A Bible Summary

Commentary:

The Ten Commandments (Law) were given to mankind to show them that they are sinners (Romans 3:20 and 7:7). The purpose of the law was not to make people righteous because no one can keep the law. No one can be justified before God by keeping the law because no one can be perfect.

Related Verses

Romans 2:12 For as many as have sinned without law will also perish without law, and as many as have sinned in the law will be judged by the law.

Romans 3:19-20 Now we know that whatever the law says, it says to those who are under the law, that every mouth may be stopped, and all the world may become guilty before God. Therefore by the deeds of the law no flesh will be justified in His sight, for by the law is the knowledge of sin.

Romans 7:7 I (Paul) would not have known sin except through the law. For I would not have known covetousness unless the law had said, "You shall not covet."

Galatians 3:1-3 O foolish Galatians! Who has bewitched you that you should not obey the truth, before whose eyes Jesus Christ was clearly portrayed among you as crucified? This only I want to learn from you: Did you receive the Spirit by the works of the law, or by the hearing of faith? Are you so foolish? Having begun in the Spirit, are you now being made perfect by the flesh (by keeping the law)?

Galatians 2:16 A man is not justified by the works of the law but by faith in Jesus Christ, even we have believed in Christ Jesus, that we might be justified by faith in Christ and not by the works

Chapter 8

of the law; for by the works of the law no flesh shall be justified (you will go to heaven by believing in Christ, not because you are good).

Galatians 3:10-13 For as many as are of the works of the law are under a curse; for it is written, "Cursed is everyone who does not continue in all things which are written in the book of the law, to do them." But that no one is justified by the law in the sight of God is evident, for "The just shall live by faith." Yet the law is not of faith; but, "The man who does them shall live by them." Christ has redeemed us from the curse of the law, having become a curse for us (for it is written, "Cursed is everyone who hangs on a tree").

Galatians 3:24-25 Therefore the law has become our tutor to bring us to Christ, that we might be justified by faith. But after faith has come, we are no longer under a tutor (the saved are not under the law).

Galatians 5:4 & 18 You have been estranged from Christ, you who attempt to be justified by law; you have fallen from grace. But if you are led by the Spirit, you are not under the law.

Romans 2:14-15 For when Gentiles, who do not have the law, by nature do the things contained in the law, these, although not having the law, are a law to themselves, who show the work of the law written in their hearts, their conscience also bearing witness, and between themselves their thoughts accusing or else excusing them (people know right from wrong even without knowing the law).

Hebrews 10:1 & 11:8 For the law, having a shadow of the good things to come, and not the very image of things, can never by the same sacrifices, which they offer continually year by year, make

A Bible Summary

those who approach perfect. HEBREWS 11:8 But without faith (faith in Jesus Christ) it is impossible to please Him (God).

1 Samuel 16:7 For the LORD does not see as man sees; for man looks at the outward appearance, but the LORD looks at the heart.

James 2:10 For whoever shall keep the whole law, and yet stumble in one point, he is guilty of all.

Colossians 2:13-14 And you, being dead in your trespasses and the uncircumcision of your flesh, He has made alive together with Him, having forgiven you all trespasses, having wiped out the handwritings of requirements (Mosaic Law) that was against us, which was contrary to us. And He has taken it out of the way, having nailed it to the cross.

Leviticus 18:21 & 20:2-5 You shall not let any of your descendants (seed, offspring, fetus) pass through the fire (to be killed) to Molech (sacrificed to a false god). Whoever ... gives any of his descendants to Molech, he (she) shall surely be put to death. And if the people of the land should in any way hide their eyes from the man (or woman), when he gives some of his descendants to Molech, and they do not kill him (or her), then I will set My face against that man and against his family.

Chapter 8

What Is Sin?

Romans 3:23 For all have sinned and fall short of the glory of God.

Romans 5:12-14 & 19 Therefore, just as through one man (Adam) sin entered the world, and death (separation from God) through sin, and thus death spread to all men, because all sinned (all are descendants of Adam). For until the law sin was in the world, but sin is not imputed when there is no law. Nevertheless death reigned from Adam to Moses, even over those who had not sinned according to the likeness of the transgression of Adam, who is a type of Him who was to come. For as by one man's (Adam's) disobedience many (all the world) were made sinners, so also by one Man's (Jesus') obedience many (all who believe) will be made righteous.

Romans 6:23 For the wages of sin is death (being spiritually separated from God), but the gift of God is eternal life in Christ Jesus our Lord.

1 John 1:8-9 If we say that we have no sin, we deceive ourselves, and the truth is not in us. If we confess our sins, He is faithful and just to forgive us our sins and to cleanse us from all unrighteousness.

Matthew 5:27-28 You have heard that it was said, "You shall not commit adultery"; but I (Jesus) say to you, that everyone who looks on a woman to lust for her has committed adultery with her already in his heart.

Matthew 5:21-22 You have heard that it was said to those of old, "You shall not murder" and whoever murders will be in danger of the judgment. "But I (Jesus) say to you that whoever is angry with his brother without a cause shall be in danger of the

A Bible Summary

judgment. And whoever says to his brother, "Raca" (probably means "empty-head"), shall be in danger of the council. But whoever says, "You fool", shall be in danger of hell fire.

Isaiah 59:1-2 Behold, the LORD's hand is not shortened, that it cannot save; nor is His ear heavy, that it cannot hear. But your iniquities (sins) have separated you from your God.

Romans 14:23 Whatever is not from faith is sin.

Commentary:

We are not sinners because we sin; rather, we sin because we are sinners. All people are born with the sinful nature because we are all descendants of Adam. We are all born without the Spirit of God and separated from God. For the wages of sin is death (eternal separation from God). All our righteous deeds are like a filthy garment.

Related Versus

Romans 6:23 For the wages of sin [is] death, but the gift of God [is] eternal life in Christ Jesus our Lord.

Isaiah 64:6 But we are all like an unclean [thing], And all our righteousnesses [are] like filthy rags; We all fade as a leaf, And our iniquities, like the wind, Have taken us away.

Chapter 8

Who Is Jesus?

John 10:30 I and My Father are one.

John 1:1-3 In the beginning was the Word (Jesus), and the Word was with God, and the Word (Jesus) was God. He was in the beginning with God. All things were made through Him, and without Him nothing was made that was made.

John 1:14 The Word (Jesus) became flesh, and dwelt among us, and we beheld His glory, the glory as of the only begotten from the Father, full of grace and truth.

Colossians 1:15-16 He (Jesus) is the image of the invisible God, the firstborn over all creation. For by Him (Jesus) all things were created that are in heaven and that are on earth, visible and invisible, whether thrones or dominions or principalities or powers. All things were created through Him and for Him.

Colossians 2:9-10 For in Him (Jesus) dwells all the fullness of the Godhead bodily; and you are complete in Him, who is the head of all principality and power.

Philippians 2:9-11 Therefore God also highly exalted Him and given Him the name which is above every name, that at the name of Jesus every knee should bow, of those who are in heaven, and of those on earth, and of those under the earth, and that every tongue should confess that Jesus Christ is Lord, to the glory of God the Father.

John 11:25 Jesus said to her, "I am the resurrection and the life. He who believes in Me, though he may die, he shall live. And whoever lives and believes in Me shall never die."

John 14:6-7 Jesus said to him, "I am the way, and the truth, and the life. No one comes to the Father, except through Me. If you

A Bible Summary

had known Me, you would have known My Father also; and from now on you know Him and have seen Him."

Mark 14:62 And Jesus said, "I AM. And you will see the Son of Man sitting at the right hand of Power, and coming with the clouds of heaven." JOHN 8:58 Jesus said to them, "Most assuredly, I say to you, before Abraham was, I AM."

1 Timothy 3:16 And without controversy great is the mystery of godliness: God was manifested in the flesh, justified in the Spirit, seen by angels, preached among the Gentiles, believed on in the world, received up in glory. (Jesus was God in the flesh in the likeness of a man.)

Ephesians 3:9 The mystery, which from the beginning of the ages has been hidden in God who created all things through Jesus Christ. (Jesus was the Creator.)

1 Timothy 1:15-17 This is a faithful saying and worthy of all acceptance, that Christ Jesus came into the world to save sinners, of whom I (Paul) am chief. However, for this reason I obtained mercy, that in me first Jesus Christ might show all longsuffering, as a pattern to those who are going to believe in Him for everlasting life. Now to the King eternal, immortal, invisible, to God who alone is wise, be honor and glory forever and ever. Amen.

Matthew 1:23 "Behold, a virgin shall be with child, and shall bear a Son, and they shall call his name Immanuel," which translated means, "God with us."

John 20:27-29 Then He (Jesus) said to Thomas, "Reach your finger here, and look at My hands; and reach your hand here, and put it into My side. Do not be unbelieving, but believing." And Thomas answered and said to Him, "My Lord and my God!" Jesus

Chapter 8

said to him, "Thomas, because you have seen Me, you have believed. Blessed are those who have not seen and yet have believed."

Isaiah 43:11-13 "I (God), even I, am the LORD, And besides Me there is no savior. I have declared and saved, I have proclaimed, And there was no foreign god among you; Therefore you are My witnesses," Says the LORD, "that I am God. Indeed before the day was, I am He; And there is no one who can deliver out of My hand; I work, and who will reverse it?"

Commentary:

Jesus Christ was Himself God manifested in the flesh. God chose to become a man in order to redeem us from our sin (John 1:1-14). Jesus Christ was the Creator (John 1:3, Col. 1:15-16, Ephesians 3:9, and Hebrews 1:2). God exists in three persons: God the Father, God the Son (Jesus Christ), and God the Holy Spirit. When Jesus said, "I AM," he was equating Himself with the God of The Old Testament. The Pharisees accused Jesus of blasphemy (Matthew 26:64-65) because he claimed to be God.

In 1 Timothy 1:17, Paul simply could go no further without sounding out this tremendous doxology. Who is "the King eternal?" He is the Lord Jesus Christ. And who is the Lord Jesus? He is "the only wise God." Paul is teaching us that the Lord Jesus was God. Paul considered Him to be God manifested in the flesh, and here he gives this wonderful testimony to that fact.

1 Timothy 1:15 says, "Christ Jesus came into the world to save sinners," and in Isaiah 43:11, God says, "besides Me there is no savior." You cannot believe that Jesus is the Savior unless you also believe that Jesus Christ is God in the flesh. Also see John 4:25-26, 5:18, 8-58, 9-35, 14:9-11; Hebrews 1:3, 8-9; Philippians 2:6; Titus 2:13; 2 Peter 1:1; Romans 9:5; 1 John 5:20; Isaiah 9:6,

A Bible Summary

44:6, 45:21; Revelation 22:8-9; Luke 24:52; and Mark 2:5-10. Jesus has all of the attributes of God because He is God.

Jesus Was Crucified for the Sins of the Whole World

Jesus Was Resurrected Bodily from the Tomb

Chapter 8

Salvation Is by Grace through Faith.

Romans 10:9-10 If you confess with your mouth the Lord Jesus and believe in your heart that God has raised Him from the dead, you will be saved. For with the heart one believes to righteousness, and with the mouth confession is made to salvation.

1 Corinthians 15:3-4 Christ died for our sins according to the Scriptures, and that He was buried, and that He rose again the third day according to the Scriptures.

Romans 1:16 I am not ashamed of the gospel of Christ, for it is the power of God to salvation for everyone who believes.

Acts 4:12 "Nor is there salvation in any other, for there is no other name under heaven given among men by which we must be saved."

1 John 5:11-13 And this is the testimony: that God has given us eternal life, and this life is in His Son. He who has the Son has life; he who does not have the Son of God does not have life. These things I have written to you who believe in the name of the Son of God, that you may know that you have eternal life, and that you may continue to believe in the name of the Son of God.

John 3:16-18 For God so loved the world that He gave His only begotten Son, that whoever believes in Him should not perish but have everlasting life. For God did not send His Son into the world to condemn the world, but that the world through Him might be saved. He who believes in Him is not condemned; but he who does not believe is condemned already, because he has not believed in the name of the only begotten Son of God.

A Bible Summary

John 3:36 "He who believes in the Son has everlasting life; and he who does not believe the Son shall not see life, but the wrath of God abides on him."

Ephesians 2:8-9 For by grace you have been saved through faith, and that not of yourselves; it is the gift of God; not of works, lest anyone should boast.

John 3:3 & 6-7 "Most assuredly, I say to you, unless one is born again (born spiritually), he cannot see the kingdom of God. That which is born of the flesh is flesh; and that which is born of the Spirit is spirit. You must be born again."

2 Corinthians 5:17 If anyone is in Christ, he is a new creation; old things have passed away; behold, all things have become new.

Romans 5:10 For if when we were enemies we were reconciled to God through the death of His Son, much more, having been reconciled, we shall be saved by His life.

1 Corinthians 1:18 For the message of the cross is foolishness to those who are perishing, but to us who are being saved it is the power of God (unto salvation).

Romans 6:14 For sin shall not have dominion over you, for you are not under law but under grace.

Romans 7:6 But now we have been delivered from the law, having died to what we were held by, so that we should serve in the newness of the Spirit and not in the oldness of the letter (Law).

Romans 8:1-2 There is therefore now no condemnation for those who are in Christ Jesus, who do not walk according to the

Chapter 8

flesh, but according to the spirit. For the law of the Spirit of life in Christ Jesus has made me free from the law of sin and death.

1 Corinthians 6:12 & 19 All things are lawful for me, but all things are not helpful. All things are lawful for me, but I will not be brought under the power of any. Or do you not know that your body is a temple of the Holy Spirit who is in you.

2 Corinthians 3:5-6 Not that we are sufficient of ourselves to think of anything as being from ourselves, but our sufficiency is from God, who also made us sufficient as ministers of the new covenant, not of the letter (the Law) but of the Spirit; for the letter kills, but the spirit gives life.

1 Peter 2:24 He Himself bore our sins in His own body on the tree, that we, having died to sins, might live for righteousness - by whose stripes you were healed.

Romans 5:8 But God demonstrates His own love toward us, in that while we were still sinners, Christ died for us.

Titus 3:5-6 (God has saved us) not by works of righteousness which we have done, but according to His mercy He saved us, through the washing of regeneration and renewing of the Holy Spirit, whom He poured out on us abundantly through Jesus Christ our Savior.

John 1:12-13 But as many as received Him (Jesus), to them He gave the right to become children of God, to those who believe in His name: who were born, not of blood, nor of the will of the flesh, nor of the will of man, but of God.

Galatians 2:20 I have been crucified with Christ; it is no longer I who live, but Christ lives in me; and the life which I now live in

A Bible Summary

the flesh I live by faith in the Son of God, who loved me and gave Himself for me.

Colossians 1:27 The mystery - which is Christ in you, the hope of glory.

Matthew 9:17 "Nor do people put new wine into old wineskins, or else the wineskins break, the wine is spilled, and the wineskins are ruined. But they put new wine into new wineskins, and both are preserved. (You cannot mix Law and Grace; you are either condemned under the Law or redeemed by the Grace of God.)

Romans 16:25-26 Now to Him (God) who is able to establish you according to my gospel and the preaching of Jesus Christ, according to the revelation of the mystery which was kept secret since the world began but now .... has been made known to all nations, ... for obedience to the faith.

Ephesians 1:7 In Him we have redemption through His blood, the forgiveness of sins, according to the riches of His grace.

Hebrews 1:3 (Jesus Christ) who being the brightness of His glory and the express image of His person, and upholding all things by the word of His power, when He had by Himself purged our sins, sat down at the right hand of the Majesty on high.

Hebrews 2:17 Therefore, in all things He had to be made like His brethren, that He might be a merciful and faithful High Priest in things pertaining to God, to make propitiation for the sins of the people (appeasing or removing their sins).

Hebrews 9:27 And as it is appointed for men to die once, but after this the judgment, 28 so Christ was offered once to bear

Chapter 8

the sins of many. To those who eagerly wait for Him He will appear a second time, apart from sin, for salvation.

1 John 1:7 But if we walk in the light as He is in the light, we have fellowship with one another and the blood of Jesus Christ His Son cleanses us from all sin.

Revelation 1:5-6 and from Jesus Christ, the faithful witness, the firstborn from the dead, and the ruler over the kings of the earth. To Him who loved us and washed us from our sins in His own blood, and has made us kings and priests to His God and Father, to Him be glory and dominion forever and ever. Amen.

Revelation 20:15 And anyone not found written in the Book of Life was cast into the lake of fire.

Revelation 21:27 But there shall by no means enter it anything that defiles, or causes an abomination or a lie, but only those who are written in the Lamb's Book of Life.

Commentary:

Every person will someday stand before the judgment seat of God (Hebrews 9:27). He will stand either condemned or pardoned having believed God's plan of salvation that Jesus Christ died for our sins and was raised again according to the Scriptures. Every person will confess Jesus Christ as Lord and Savior. For many it will be too late. Those who did not confess Jesus Christ as Lord in their Earthly life will be cast into the eternal Lake of Fire. Those who have confessed Jesus Christ as Lord will have their names written in the Lamb's Book of Life and will enjoy eternity in Heaven with God. (Revelation 21:27)

A Bible Summary

We Are Not Justified by Our Good Works!

John 6:28-29 Then they said to Him, "What shall we do, that we may work the works of God?" Jesus answered and said to them, "This is the work of God, that you believe in Him (Himself) whom He (God) sent."

Luke 17:10 When you have done all those things which you are commanded, say, "We are unprofitable (unworthy) servants (to God). We have done what was our duty to do."

James 2:18-26 But someone will say, "You have faith, and I have works." Show me your faith without your works, and I will show you my faith by my works. For as the body without the spirit is dead, so faith without works is dead also.

1 Corinthians 3:13-15 Each one's work will become manifest; for the Day will declare it, because it will be revealed by fire; and the fire will test each one's work, of what sort it is. If anyone's work is burned, he will suffer loss; but he himself will be saved, yet so as through fire.

Commentary:

Jesus said the work of God was belief in Him, not fleshly works. When a person has faith, it will result in good works which do not require recognition from others. God will test all good works with fire, and a person's works done for his own reward or out of pride will be burned up. James was an Apostle to the Jews who were still under the Old Testament doctrine that required both faith and works. The Apostle Paul declared that works are not required in this Dispensation of Grace.

Chapter 8

Who Is the Indwelling Holy Spirit?

Psalms 51:11 Do not cast me away from Your presence, And do not take Your Holy Spirit from me. 12 Restore to me the joy of Your salvation, And uphold me [by] [Your] generous Spirit. 13 [Then] I will teach transgressors Your ways, And sinners shall be converted to You. 14 Deliver me from the guilt of bloodshed, O God, The God of my salvation, [And] my tongue shall sing aloud of Your righteousness. 15 O Lord, open my lips, And my mouth shall show forth Your praise.

Commentary:

God gives the indwelling Holy Spirit to all who believe the gospel in the current Dispensation of Grace. The indwelling of the Holy Spirit did not occur in any of the previous dispensations. Certainly the Holy Spirit came into the lives of many of God's people, but this was on a temporary basis. King David pleaded with God not to take the Holy Spirit from him. King David's example shows that a permanent indwelling was never given.

Luke 1:13 But the angel said to him, "Do not be afraid, Zacharias, for your prayer is heard; and your wife Elizabeth will bear you a son, and you shall call his name John. 14 "And you will have joy and gladness, and many will rejoice at his birth. 15 "For he will be great in the sight of the Lord, and shall drink neither wine nor strong drink. He will also be filled with the Holy Spirit, even from his mother's womb.

Commentary:

John the Baptist was filled with the Holy Spirit from his mother's womb. We can assume that his filling continued.

A Bible Summary

John 15:26 But when the Helper comes, whom I shall send to you from the Father, the Spirit of truth who proceeds from the Father, He will testify of Me.

John 14:16-18 And I will pray the Father, and He will give you another Helper , that He may abide with you forever-- 17 "the Spirit of truth, whom the world cannot receive, because it neither sees Him nor knows Him; but you know Him, for He dwells with you and will be in you. 18 "I will not leave you orphans; I will come to you.

Commentary:

Jesus told his disciples that He was going away (death and resurrection), but He would send a Helper. We know from Scripture that the Helper was the gift of the indwelling Holy Spirit given to believers at the moment of salvation.

Luke 24:1-9 Now on the first [day] of the week, very early in the morning, they, and certain [other] [women] with them, came to the tomb bringing the spices which they had prepared. 2 But they found the stone rolled away from the tomb. 3 Then they went in and did not find the body of the Lord Jesus. 4 And it happened, as they were greatly perplexed about this, that behold, two men stood by them in shining garments. 5 Then, as they were afraid and bowed [their] faces to the earth, they said to them, "Why do you seek the living among the dead? 6 "He is not here, but is risen! Remember how He spoke to you when He was still in Galilee, 7 "saying, `The Son of Man must be delivered into the hands of sinful men, and be crucified, and the third day rise again.' " 8 And they remembered His words. 9 Then they returned from the tomb and told all these things to the eleven and to all the rest.

Chapter 8

John 20:1 Now on the first [day] of the week Mary Magdalene went to the tomb early, while it was still dark, and saw [that] the stone had been taken away from the tomb. 2 Then she ran and came to Simon Peter, and to the other disciple, whom Jesus loved, and said to them, "They have taken away the Lord out of the tomb, and we do not know where they have laid Him." 3 Peter therefore went out, and the other disciple, and were going to the tomb. 4 So they both ran together, and the other disciple outran Peter and came to the tomb first. 5 And he, stooping down and looking in, saw the linen cloths lying [there]; yet he did not go in. 6 Then Simon Peter came, following him, and went into the tomb; and he saw the linen cloths lying [there], 7 and the handkerchief that had been around His head, not lying with the linen cloths, but folded together in a place by itself. 8 Then the other disciple, who came to the tomb first, went in also; and he saw and believed. 9 For as yet they did not know the Scripture, that He must rise again from the dead.

John 20:19 Then, the same day at evening, being the first [day] of the week, when the doors were shut where the disciples were assembled, for fear of the Jews, Jesus came and stood in the midst, and said to them, "Peace [be] with you." 20 When He had said this, He showed them [His] hands and His side. Then the disciples were glad when they saw the Lord. 21 So Jesus said to them again, "Peace to you! As the Father has sent Me, I also send you." 22 And when He had said this, He breathed on [them], and said to them, "Receive the Holy Spirit."

Commentary:

Later on the day of His resurrection, Jesus entered through closed doors into the room where the disciples were discussing His resurrection. The women who had visited His tomb and heard the angel tell them that He had risen were explaining their

A Bible Summary

experience to the others. After Jesus entered, He breathed on them and said to them, "Receive the Holy Spirit." This was the exact fulfillment of Jesus' promise to send a Helper. It was the beginning of the indwelling of the Holy Spirit in the saints.

Acts 2:1 When the Day of Pentecost had fully come, they were all with one accord in one place. 2 And suddenly there came a sound from heaven, as of a rushing mighty wind, and it filled the whole house where they were sitting. 3 Then there appeared to them divided tongues, as of fire, and [one] sat upon each of them. 4 And they were all filled with the Holy Spirit and began to speak with other tongues, as the Spirit gave them utterance. 5 And there were dwelling in Jerusalem Jews, devout men, from every nation under heaven. 6 And when this sound occurred, the multitude came together, and were confused, because everyone heard them speak in his own language. 7 Then they were all amazed and marveled, saying to one another, "Look, are not all these who speak Galileans? 8 "And how [is] [it] [that] we hear, each in our own language in which we were born?

Commentary:

The people did not speak in other tongues on Resurrection Day when Jesus gave them the Holy Spirit. The revealing of the indwelling Holy Spirit with great power and astonishment occurred in Acts 2, but this was after Jesus breathed on the disciples and gave them the Holy Spirit. Pentecost was the revealing of the power and majesty of the Holy Spirit. It was not the beginning of the indwelling Holy Spirit in the Saints. The Holy Spirit was given in John 20:22.

Speaking in tongues and performing healing miracles are not evidences of salvation or the indwelling Holy Spirit. Satan and his agents can do those things. Rightly dividing the Word of Truth is

Chapter 8

evidence of possession of the Holy Spirit. Satan and his agents always twist Scripture.

The Holy Spirit was also revealed to the believing uncircumcised Gentiles, who were not bound by the Law or circumcision. The indwelling Holy Spirit was a gift from God as Jesus had promised.

What Is the Rapture of the Church?

1 Thessalonians 4:16-17 For the Lord Himself will descend from heaven with a shout, with the voice of an archangel, and with the trumpet of God. And the dead in Christ will rise first. 17 Then we who are alive [and] remain shall be caught up together with them in the clouds to meet the Lord in the air. And thus we shall always be with the Lord.

A Bible Summary

Repent and Follow Jesus Christ.

Matthew 4:17 From that time Jesus began to preach and to say, "Repent, for the kingdom of heaven is at hand."

Luke 13:3 Unless you repent you will all likewise perish. LUKE 24:47 Jesus said, Repentance and remission of sins should be preached in His name to all nations, beginning at Jerusalem."

Acts 2:38 Then Peter said to them, "Repent, and let every one of you be baptized in the name of Jesus Christ for the remission of sins; and you shall receive the gift of the Holy Spirit."

Acts 20:21 Paul said to them, "Testify to Jews, and also to Greeks, repentance toward God and faith toward our Lord Jesus Christ."

Hebrews 5:9 And having been perfected, He (Jesus) became the author of eternal salvation to all who obey Him.

Commentary:

Jesus continually called upon the people to follow Him and believe that He was God in human form, the great I AM, their Savior. People who know about Jesus but do not obey Him do not have salvation. They deceive themselves.

Chapter 8

What Must I Do?

The Bible tells us to confess to God with our mouth our faith in the Lord Jesus. This means to pray to God according to the gospel of Jesus Christ found in Romans 10:9-10.

> **Romans 10:9** If you confess with your mouth the Lord Jesus and believe in your heart that God has raised Him from the dead, you will be saved. 10 For with the heart one believes to righteousness, and with the mouth confession is made to salvation.

You can have everlasting life with God (salvation) only by your faith in Jesus Christ. Salvation is by grace through faith based upon the redemptive work of our Lord Jesus Christ. Jesus took upon Himself the sins of all mankind by His death on the cross, and His resurrection from the grave provides salvation to all who believe. His resurrection gives us eternal life. All who receive the Lord Jesus Christ through faith are born again of the Holy Spirit and thereby become the children of God by the divine baptism of the Holy Spirit into the Body of Christ. God knows your heart. You can talk to God. Tell Him that you believe and trust in what Jesus Christ has done for you. Pray to God from your heart.

A Bible Summary

Salvation Prayer

Dear God,

I know that I have done many sinful deeds. I know that no one can be justified before You by keeping commandments or doing good works because no one can be perfect. I believe that You came to Earth as a man, Jesus Christ. I believe Jesus took upon Himself the sins of the whole world when He was crucified and physically died on a cruel cross at Calvary. I believe Jesus Christ was buried. I believe in my heart that on the third day You, God, physically raised Jesus bodily from the dead. I believe Jesus ascended to Heaven and sat down at Your right hand where He reigns as my Advocate.

I believe that You, God the Father, Jesus Christ the Son, and the Holy Spirit are one God. I thank Jesus for taking away my sins forever. I believe Jesus Christ is my Lord and Savior. I believe You will send the Holy Spirit to live within me. I believe I am born again in the Spirit and will now have eternal life with You. Thank You, God, for forgiving me and giving me everlasting life.

In the name of Jesus Christ I pray, Amen.

I want to record the date that I prayed to confess with my mouth the Lord Jesus and believed in my heart that God raised Him from the dead.

Name: _____ Date: _____
Permission is granted to copy, sign, and keep this page.

Index

A

acetoacetic acid 17, 30
acetone 17, 30
acupuncture 57
adipose tissue 17, 21
adrenal gland 18, 19, 21
adrenaline 3, 11, 15, 20, 22, 23, 24, 89, 286
Africa 38, 98, 189, 261
African 38, 39, 40, 43, 295
Allan, Christian B., Ph.D. 84, 115
allergies 10, 19, 95, 195, 204, 228, 236, 270, 278
almond 125, 249, 271, 272
alpha-linolenic fatty acid (ALA) 36, 218
Alzheimer's disease 16
amino acids 3, 7, 10, 18, 32, 36, 52, 56, 63, 88, 89, 101, 103, 113, 114, 124, 148, 163, 216, 217, 222, 238, 246, 251, 253, 254, 272
Amundsen, Roald 66
Anabolic Diet 85
Anderson
 Karsen xiii, 42, 65, 89, 99, 266, 285
antibiotic 205, 212, 215, 224, 226, 227, 230, 233, 237
anti-inflammatory 18, 214, 276, 278
anti-matter 321
anus 44, 235
Apollo 1 290
appendix 53, 224
apricot 125, 249, 271
arachidonic acid 125, 126, 159, 200, 218, 219
asthma v, xxxii, 48, 168, 169, 170, 195, 196
atherosclerosis 12, 113
atherosclerotic plaque 221
Atkins
 Dr. Robert C. x, xiii, 13, 33, 45, 61, 74, 76, 90, 120, 127, 132, 165, 236, 250, 256, 271, 297
attention deficit hyperactivity disorder (ADHD) 9, 24, 25
autism xxxvi, 207
autoimmune disease xx, xxxii, xxxiii, xxxvii, xxxviii, 48, 127, 170, 173, 176, 182, 192, 194, 199, 233, 250, 264, 271
autoimmune diseases xxii, xxvi, xxxii, 7, 60, 71, 251, 264, 270, 276, 278
avocado 111, 243, 267

B

baby xxxvi, 9, 165, 295
baby foods 9
bacteria xx, xxiv, xxxi, 41, 43, 47, 48, 49, 51, 52, 53, 54, 55, 57, 58, 84, 91, 96, 98, 99, 101, 106, 139, 174, 175, 190, 201, 206, 212, 214, 222, 223, 224, 225, 226, 227, 228, 229, 232, 234, 235, 238, 245, 247, 249, 252, 255, 261, 262, 263, 266, 267, 268, 269, 271, 272, 274, 306, 334
Ball jar 147
Banting, William xiii, 5, 132
bear 42, 43, 295, 382, 389, 395, 398
beef 7, 64, 79, 86, 101, 102, 107, 112, 113, 120, 177, 240, 243, 246, 266, 274, 281, 305
Bernstein, Dr. Richard K. 112, 149
beta cells 12
beta-hydroxybutyric acid 17, 30
Bible Summary v, 373
Big Bang iii, v, xvii, xix, xxv, xxix, xxx, xxxi, 282, 292, 299, 301, 302, 308, 313, 314, 315, 316, 317, 318, 319, 320, 321, 322, 324, 326, 331, 332, 338, 341, 342, 343, 344, 346, 347, 348, 349, 350, 351, 352, 353, 354, 356,

Index

357, 358, 359, 361, 362, 365, 367, 369, • 371
Big Bang Theory   xix, xxv, xxx, xxxi, 314, 315, 316, 317, 318, 319, 324, 332, 341, 343, 348, 350, 351, 352, 354, 357, 359, 361, 365, 367, 371
Big Squirt Theory   342
biking   23
biosynthesis   218
black hole   333, 338, 346, 347, 348, 350, 366
Black Hole Theory   346
blood   x, xxxv, xxxvi, 2, 3, 9, 10, 11, 12, 14, 15, 17, 18, 19, 20, 21, 23, 24, 30, 32, 33, 34, 36, 40, 49, 51, 65, 69, 82, 84, 86, 90, 93, 94, 103, 109, 110, 112, 115, 119, 121, 132, 133, 136, 147, 173, 174, 194, 197, 199, 217, 220, 245, 247, 250, 251, 254, 261, 262, 266, 267, 270, 376, 394, 395, 396
blood type   128, 129, 130
body fat   6, 11, 12, 13, 17, 24, 30, 109, 249
bodybuilding   23
bone 8, 19, 20, 24, 36, 43, 55, 65, 88, 91, 98, 113, 114, 116, 117, 163, 216, 217, 218, 231, 243, 248, 254, 272, 273
borage oil   127, 250, 268, 271
Boulder   75
bowel diseases   xxxii, 41, 80, 87, 95, 96, 99, 185, 215, 223, 233, 238, 261
brainwashing   xi, xii, xv, xviii, xx, xxi, xxii, xxv, xxvii, xxxi, xxxii, xxxviii, 33, 67, 76, 88, 91, 118, 176, 284, 286, 287, 288, 289, 290, 292, 293, 294, 295, 296, 297, 298, 299, 300, 301, 302, 303, 304, 305, 310, 311, 314, 334, 337, 343, 372
bread   38, 39, 43, 76, 92, 98, 120, 137, 139, 201, 202, 237, 246, 272, 273, 275

breast milk   9, 65
Burkitt
   Dr. Dennis 38, 40, 43, 55, 57, 98, 144, 261
butter   69, 74, 75, 76, 77, 79, 86, 120, 152, 179, 243, 246, 248, 249, 263, 266, 273, 275

C

C. diff   222, 225
calcium   20, 43, 88, 98, 113, 114, 117, 250, 272
Canada   xiii, 41, 42, 64, 65, 66, 75, 118, 162, 164, 298
cancer   x, xxvi, 4, 6, 9, 14, 27, 31, 40, 41, 43, 44, 45, 50, 54, 55, 56, 57, 60, 61, 62, 64, 65, 69, 70, 71, 73, 75, 80, 86, 89, 93, 94, 95, 96, 99, 102, 103, 106, 112, 113, 115, 118, 121, 125, 133, 136, 140, 144, 147, 149, 159, 160, 163, 164, 165, 183, 185, 203, 209, 218, 219, 220, 229, 240, 245, 261, 263, 266, 270, 272, 273, 275, 304, 306, 307
candida albicans   204, 205
Carbohydrate Consumption Syndrome   2
Carbohydrate Death Curve   27
carbohydrates   xi, xii, xxv, xxvi, xxxv, xxxviii, 2, 3, 5, 6, 8, 9, 10, 11, 12, 13, 14, 15, 17, 18, 19, 23, 24, 25, 26, 27, 28, 31, 33, 35, 39, 42, 46, 48, 49, 50, 51, 52, 53, 54, 55, 57, 58, 60, 65, 68, 69, 72, 76, 77, 80, 83, 84, 85, 86, 87, 90, 91, 96, 97, 98, 101, 102, 103, 115, 118, 119, 120, 140, 148, 150, 152, 157, 162, 164, 190, 201, 202, 204, 205, 214, 221, 222, 223, 224, 226, 227, 228, 233, 235, 238, 239, 245, 246, 247, 249, 252, 253, 261, 263, 264, 267, 269, 270, 271, 273, 274, 275, 292, 301, 304, 306, 308

Index

carcinogens 213
cardiologist xiii, xxxv, 45, 77, 120
caribou 43, 63, 64, 65, 266
Carlson's Lemon Flavored Cod Liver Oil
 82, 126, 249
Carlson's® 108, 256, 267, 295
carnivorous 8, 58, 80, 101, 215
catabolic state 20
catabolize 18
cataract 15
cats 58, 215
Celiac disease 186
cereal 57, 87, 92, 144, 153, 155, 156, 157, 162, 171, 237
cheese 35, 69, 76, 77, 177, 179, 243, 247, 250, 254, 263, 265, 267, 269, 273, 281
Chernobyl Nuclear Power Station 291
chicken 78, 107, 108, 153, 224, 246
chimpanzees 294
Chinese medicine 57, 170, 306
cholesterol xxxv, xxxvi, 14, 23, 69, 70, 72, 86, 90, 104, 109, 110, 113, 120, 121, 163, 167, 194, 238, 247, 249, 267, 273, 306
Cipro® 212
Clostridium difficile 52, 222, 225
Coca-Cola® 150
coconut oil 76, 77, 78, 120, 238, 243, 246, 247, 248, 249, 264, 266, 267, 269, 273, 275, 281
coffee 268, 269, 286
collagen 8, 19, 36, 88, 89, 91, 216, 217, 254
collagen matrix 8, 88, 89
colon xii, xx, xxxii, xxxiv, 7, 40, 41, 43, 44, 45, 46, 48, 49, 50, 51, 52, 53, 54, 55, 56, 58, 84, 96, 98, 99, 102, 112, 139, 175, 178, 185, 188, 189, 201, 202, 203, 204, 215, 222, 223, 224, 225, 226, 227, 228, 233, 235, 254, 262, 263, 266, 274

colonoscopy 234, 263
combine harvester 161, 162
constipation xxxii, 22, 44, 45, 46, 47, 48, 50, 53, 57, 84, 96, 97, 178, 180, 183, 185, 203, 215, 227, 232, 235, 240, 244, 245, 250, 253, 262
corn x, 71, 78, 125, 143, 150, 156, 159, 214, 215, 219, 220, 249, 270, 271, 272, 273, 274, 304
cortisol 3, 15, 18, 19, 20, 21, 87, 89, 91
cottonseed 71, 125, 159, 220, 249, 271
Crohn's disease xi, xx, xxxii, xxxiv, xxxvii, 44, 71, 95, 145, 159, 170, 173, 174, 175, 184, 185, 188, 189, 190, 201, 215, 220, 228, 236, 264, 276, 306, 308

D

Dark Energy 314, 320, 366, 367
Darwin, Charles 299
Deadly Food Pie 147
deficiencies 74, 111, 140, 207, 218, 249, 252
deficiency of protein and fats 5
degenerative disc disease 14, 19, 20, 24, 44, 55, 98, 118, 121, 156, 179, 183
diabetes xxvi, 2, 4, 10, 11, 12, 14, 15, 19, 27, 30, 36, 45, 57, 60, 61, 62, 64, 70, 76, 80, 86, 87, 90, 92, 93, 94, 95, 96, 97, 102, 103, 107, 112, 116, 118, 119, 133, 136, 139, 140, 143, 144, 147, 149, 161, 162, 164, 187, 215, 217, 228, 245, 251, 254, 261, 270, 271, 304, 306, 307
diarrhea xxxii, 44, 47, 48, 50, 54, 57, 84, 96, 97, 178, 180, 183, 184, 185, 203, 205, 215, 227, 232, 235, 240, 245, 250, 253, 255, 262
Diflucan® 205, 227, 255
dinosaurs 370
Dispensation of Grace 397, 398

Index

diverticulosis 44, 170, 175, 189
docosahexaenoic fatty acid (DHA) 36, 62, 218
dogs 8, 58, 65, 67, 102, 215, 224, 238, 271
doppler *effect* xxix, xxx, 319, 321, 333, 339
Doxycycline 225
duodenal ulcer 229
duodenum 229, 231

E

$E = M * C^2$ 315
E. coli 106, 225
Eades x, xiii
    Drs. Michael R. and Mary Dan 13, 127, 236, 250, 271
edema 16
eggs 69, 86, 107, 108, 120, 153, 157, 177, 243, 246, 281
Egypt xxvi, 14, 139
eicosapentaenoic acid (EPA) 36, 62, 218
Einstein
    Dr. Albert xix, xxx, 300, 315, 321, 338, 352, 357
Elizabeth Smart iv, 288
England xiii, 5, 38, 66, 99, 108, 132, 229
Enig, Mary G., Ph.D. 110
epinephrine 21, 53, 227
erlose 6, 270
Escherichia coli 106, 225
Eskimo xiii, xvii, xxvi, 9, 14, 31, 42, 43, 63, 64, 65, 66, 67, 89, 99, 100, 102, 115, 262, 266, 285
essential fats 3, 5, 148
essential proteins 3, 235
exercise 21, 23, 56, 86, 119, 120, 166, 192, 193, 264, 307
Expanding Universe Theory 315, 317

F

$F = (G * M_1 * M_2) / D^2$ 355
$F = M * A$ 322, 354
Fallon, Sally 110
fecal matter 50, 53, 223
feedlot 7, 240
fermentation 46, 48, 49, 51, 57, 58, 98, 124, 175, 236, 261, 274
fiber xi, xii, xx, xxvi, xxxviii, 18, 22, 38, 40, 41, 42, 43, 45, 46, 47, 48, 49, 50, 51, 52, 53, 54, 55, 57, 58, 65, 68, 78, 84, 96, 97, 98, 99, 100, 102, 144, 171, 190, 201, 202, 205, 223, 224, 226, 227, 228, 233, 235, 236, 239, 240, 245, 246, 247, 248, 249, 261, 267, 269
fibromyalgia 71, 159, 215, 220, 245, 251
fibrous foods 47
fish 3, 34, 62, 63, 65, 71, 76, 108, 124, 125, 135, 142, 160, 161, 177, 180, 185, 204, 238, 247, 248, 249, 266, 267, 274, 275, 374, 376
fissure 234
fissures 44, 175, 189, 234, 235
fistulas 175, 234, 235
Fixx, James F. 23
Flagyl® 190, 222, 225
Flat Earth Theory 343
flatulence 48, 49, 51, 84, 215, 235, 261
flour 6, 39, 64, 76, 83, 98, 120, 137, 147, 152, 161, 249, 252, 255, 269, 272
Foods That Are Absolutely Forbidden 171, 270
Foods We Should Eat 171, 266
Forbidden Foods xxxvii, 176, 201, 242
fox 43, 63
Framingham Study 70
free-range 107
French paradox 69, 76

Index

fructose 6, 8, 10, 17, 31, 93, 135, 136, 141, 143, 147, 148, 149, 150, 156, 245, 253, 261, 264, 270
fruit ix, xxxvi, 6, 8, 9, 17, 20, 25, 39, 53, 55, 57, 76, 78, 86, 87, 92, 93, 94, 96, 98, 100, 119, 120, 121, 135, 142, 146, 147, 148, 149, 150, 162, 171, 201, 203, 224, 233, 235, 237, 240, 245, 248, 261, 264, 270, 271, 275, 285, 373
fungi 7, 52, 54, 204, 222, 226, 228, 232, 235, 255, 271, 274
fungus 7, 51, 204, 205, 206, 226

G

$G = 6.672 \times 10^{-11}$ N m^2 kg^{-2} 355, 356, 357
galaxies iv, xix, xxix, xxx, 301, 314, 318, 319, 320, 321, 322, 323, 324, 327, 331, 332, 333, 334, 335, 336, 338, 339, 340, 341, 343, 344, 350, 351, 353, 354, 357, 358, 362, 364, 365, 371, 372
galaxy
    Cross 363
    M100 344
    M51 337, 363
    M81 333
    NGC2207 353
Galileo Galilei 284
gallbladder 6, 109, 121, 122, 175, 249
Gap Theory 370
garbage collection trucks 224
Gaseous Pillars M16 358, 359
gastroenterologists xi, xxxiii, 44, 45, 46, 48, 50, 53, 98, 190, 201, 227, 306
gastroesophageal reflux disease 230
General Theory of Relativity xix, xxxi, 300, 325, 329, 338, 345, 352, 356, 357
genetically modified organism (GMO) 167

Genova Diagnostics® 206, 224, 232
Georgian 75
GERD 170, 229, 230, 231
giardia 222
Gittleman, Ann Louise 163, 164
glucagon 17
glucocorticoid 18
gluconeogenesis 3, 10
glucose 3, 6, 9, 10, 11, 12, 13, 14, 15, 17, 18, 19, 20, 21, 24, 25, 30, 31, 32, 33, 34, 51, 54, 57, 65, 81, 82, 84, 86, 87, 90, 94, 115, 136, 141, 150, 220, 228, 254, 264, 270
Glucose-Galactose Malabsorption 54, 228
glutamine amino acid 35, 253
glycation 13, 15, 90, 221
goats 40, 57, 58, 247, 267, 272
GOD xxx, xxxi, 153, 177, 180, 232, 299, 302, 321, 356, 363, 364, 370, 371, 373, 374, 375, 376, 377, 378, 379, 380, 381, 382, 383, 384, 385, 386, 387, 388, 389, 390, 392, 393, 394, 395, 396, 397, 398, 402, 403, 404, 405
Graham, Sylvester 152
grain drill 161
granula 153
grapeseed 125, 249, 271
gravitational repulsion 321
gravitational wave 329, 338
grinding mill 137
ground beef 238, 243
Groves, Barry, PhD 100

H

hazelnut 125
HDL xxxv, 73, 90, 109, 120
HealthCheckUSA® 213
Hearst, Patricia 288
heart attack 12, 15, 23, 90, 110, 157

Index

heart disease x, xxvi, xxxv, 2, 4, 6, 10, 12, 13, 14, 20, 22, 23, 27, 31, 45, 57, 60, 61, 62, 64, 65, 69, 70, 71, 75, 76, 79, 80, 85, 86, 90, 93, 94, 95, 97, 103, 109, 110, 112, 113, 118, 120, 121, 127, 133, 136, 139, 140, 144, 147, 149, 157, 159, 162, 164, 165, 219, 220, 221, 228, 240, 245, 250, 251, 270, 271, 292, 304, 306, 307
heavy metals 213
Helicobacter pylori 229, 230
herbivores 8
High-Carbohydrate Foods 2, 29, 236
high-fiber diet xi, xx, xxvi, 38, 46, 47, 50, 53, 55, 96, 97, 98, 144, 201, 227, 235
hip fracture 19, 20, 92, 116
Holy Spirit 376, 381, 390, 394, 398, 399, 400, 401, 402, 403, 404, 405
homeopathic medicine 57
homocysteine 109
honey 6, 17, 39, 76, 141, 237, 248, 265, 270, 380
Hubble Space Telescope 360
Hubble, Dr. Edwin xix, xxix, 301, 314, 319, 334, 335, 336, 338, 339, 351, 357
Hudson Bay Company 43, 64
human growth hormone 24
hydrochloric acid 231
hyperactive 24
hyperadrenalemia 22
hyperglycemia 11, 15
hyperinsulinemia 15, 22, 90, 94, 136, 220
hypertension 13, 112, 174, 247, 267
hypertriglyceridemia 22
hyperuricemia 93, 94, 136
hypoglycemia 10, 12, 15, 18, 20, 22, 34, 36, 118, 217, 254, 261

I

IBD iii, xxxii, xxxiii, xxxiv, 50, 53, 170, 175, 179, 201, 202, 203, 215, 216, 224, 226, 236, 237, 242, 245, 248, 249, 254, 265, 269
immune system xi, xxxii, xxxiii, xxxiv, 19, 28, 36, 44, 56, 74, 89, 102, 113, 171, 173, 174, 175, 176, 182, 186, 188, 190, 192, 194, 195, 197, 199, 204, 212, 214, 215, 216, 217, 246, 251, 254, 264, 277, 278
Indian 9, 102
Industrial Revolution 134, 135, 137
inflammatory bowel disease 55, 70, 71, 86, 93, 136, 159, 175, 184, 188, 201, 219, 220, 223, 264
inflammatory bowel diseases 6, 14, 27, 38, 41, 43, 45, 50, 52, 53, 55, 57, 58, 173, 201, 202, 203, 204, 215, 232, 236, 240, 242, 278, 293, 306, 308
insulin 3, 9, 10, 11, 12, 13, 14, 15, 16, 17, 18, 19, 22, 24, 26, 30, 32, 34, 36, 53, 54, 69, 84, 86, 87, 91, 92, 93, 94, 112, 115, 119, 120, 136, 143, 149, 150, 187, 217, 220, 227, 228, 254, 264, 270, 271
insulin resistance 13, 17, 24, 31, 86, 94, 136, 143, 220
International Space Station iv, 340
Interstitial cystitis 233
intestinal gas 51, 247, 262, 267
intestines 49, 51, 96, 98, 101, 124, 167, 175, 183, 188, 202, 216, 223, 224, 231, 236, 252, 261
irritable bowel syndrome xxii, 71, 159, 175, 220, 264
Irritable bowel syndrome 183, 185
isomaltose 6, 270

Index

J

Jesus 363, 379, 381, 383, 385, 386, 387, 388, 389, 390, 391, 392, 393, 394, 395, 396, 397, 399, 400, 401, 402, 403, 404, 405
jogging 23, 119, 286
Johnson, Gaines 370
Jones, Reverend Jim 289
Joseph 139, 144, 330

K

Keflex® 212, 225
Kellogg, John Harvey 154
ketoacidosis 30, 31
Ketogenic Diet 80, 85
ketone bodies 17, 30, 31, 254
ketones 30, 31, 32, 82
ketosis iii, 30, 31, 33, 34, 244
Keys, Dr. Ancel Benjamin 144, 145
kojibiose 6, 270

L

LabCorp® 213
Lactobacillus Sporogenes 226, 256
Langerhans 12
lard 69, 76, 77, 86, 120
LDL xxxv, 12, 14, 15, 23, 70, 72, 90, 120, 221
leaky gut syndrome 52, 96, 100, 139, 173, 175, 182, 214, 233, 234, 235, 270
leg cramps 260
Lenard, Lane, Ph.D. 216, 231
Letter of Corpulence xiii, 5
lettuce xxxvi, 271
lions 8, 58
lipolysis 17
liver xxxv, 3, 10, 17, 18, 21, 32, 34, 74, 92, 99, 108, 109, 121, 174, 187, 188, 197, 199, 212, 218, 233, 243, 246, 250, 268, 277, 295, 296
longevity 4, 75, 77, 78, 79, 145, 156, 264
low-carbohydrate viii, x, xiii, 5, 10, 13, 14, 16, 18, 21, 24, 28, 30, 31, 32, 33, 34, 35, 39, 45, 49, 51, 54, 56, 61, 63, 65, 70, 72, 76, 83, 84, 85, 91, 93, 96, 97, 100, 120, 128, 142, 165, 173, 204, 228, 231, 236, 242, 244, 245, 250, 252, 253, 254, 255, 261, 264, 274
Lowell, Percival 300
lupus xx, xxxvii, 7, 27, 71, 145, 159, 170, 173, 175, 187, 195, 197, 201, 215, 220, 242, 264, 272
Lutz, Wolfgang, M.D. 84, 115

M

macronutrients 8, 58
macrophages 15, 90, 159
macular degenerative 15
magnesium 55, 98, 250, 251, 262
malabsorption 93, 253
maltose 6, 270
maltulose 6, 270
marathon 23, 56
Mars iv, xvii, xxvii, xxviii, xxix, xxxi, 292, 300, 301
Martians 300, 301
Masai 40
Mastica 229
McGovern
  Senator George 62, 163, 164, 165
meat xiii, 3, 8, 9, 25, 34, 39, 40, 42, 43, 45, 55, 56, 63, 64, 65, 66, 67, 69, 70, 73, 74, 75, 78, 80, 83, 85, 86, 87, 88, 89, 100, 101, 102, 103, 104, 106, 113, 114, 116, 117, 119, 124, 125, 126, 135, 142, 152, 153, 154, 157, 161, 167, 175, 177, 178, 179, 200, 203, 204, 205, 212, 215, 216, 219,

412

Index

231, 233, 238, 239, 240, 241, 246, 247, 248, 249, 261, 262, 266, 267, 281, 285, 286, 293, 303, 305
Medical Disclaimer viii, 172
Mediterranean Diet 72, 76, 94
metabolic type 128, 129
metabolism 3, 9, 18, 24, 81, 115, 119, 206, 264, 268
Micro Black Holes 366
Milky Way Galaxy xxx, 340, 366
minerals 6, 8, 35, 39, 40, 55, 62, 63, 81, 88, 98, 103, 111, 114, 124, 142, 148, 163, 167, 179, 207, 239, 243, 245, 246, 254, 255, 268, 305
Mitchell, Brian David 287, 288
mitochondria 10
monounsaturated 68, 71, 72, 158, 159, 160, 201, 239
MRSA bacteria 241
Mt. Palomar Observatory 345
mucosa villi 52, 222
Multidophilus® 226, 252, 257
multiple sclerosis xxxvii, 7, 27, 71, 145, 159, 170, 173, 175, 192, 194, 201, 215, 220, 242, 264, 272
mummies xxvi, 97, 139
mycotoxins 7, 204

N

NASA iv, xxvii, xxviii, xxxi, 290, 291, 323, 328, 330, 331, 337, 346, 347, 360, 365, 370
natural killer (NK) cells 159
negative mass 321, 340
North West Company 43, 64
Northwest Passage 66
NSAIDS 214, 276
nutritional quagmire 132

O

obesity 4, 6, 10, 11, 18, 61, 64, 70, 74, 86, 87, 92, 93, 94, 102, 118, 129, 136, 140, 147, 157, 161, 162, 163, 164, 215, 216, 228, 304
obsessive compulsive controlling disorder (OCCD) 311
obsessive compulsive debating disorder (OCDD) 312
Okinawa 77, 78, 79
olive oil 72, 76, 95, 160, 249, 275
omega-3 eggs 107
omega-3 fatty acids 34, 36, 39, 62, 107, 126, 127, 218, 246, 250, 267, 268
omega-6 fatty acids 7, 71, 80, 126, 127, 159, 220, 221, 238, 249, 250, 267, 269, 271, 273, 275
omega-6 vegetable oils 34, 76, 78, 103, 125, 126, 268, 304
omega-9 fat 72
optic neuritis 15
organic ix, 105, 106, 108, 123, 167, 237
osteoporosis 8, 19, 27, 36, 44, 55, 62, 86, 91, 97, 98, 113, 114, 133, 139, 140, 165, 217, 245, 251, 254, 272, 273

P

Paleo 6, 270
pancreatic enzymes 49
panose 6, 270
parasites 51, 175, 206, 232
peanut 71, 125, 159, 219, 220, 249, 271, 274
pee away bones 19, 91
pemmican 43, 64, 66, 101, 266
perfect poop 49, 262
pets in the home 224
phytoestrogens 124, 167
Plavix® xxxvi
polar bear 63, 65, 295, 296, 297

Index

polysaccharide 15, 90, 220
polyunsaturated 68, 71, 72, 73, 125, 127, 144, 158, 159, 160, 219, 220, 239, 249, 271, 274
poppy seed 125
pork 72, 76, 77, 78, 79, 86, 101, 102, 120, 160, 177, 246, 248, 266, 281, 293
Price
    Dr. Weston A. xiii, 40, 41, 74, 75, 91, 99
primrose oil 127, 250, 271
Pritikin, Nathan 62, 163, 164, 165
probiotics 101, 190, 207, 226, 245, 252, 255, 262, 272
proctitis 170, 234
professional nutritionists xxxviii, 9, 16, 33, 94, 175, 202, 307
pro-inflammatory 7, 71, 123, 126, 159, 219, 220
prostaglandin 127, 159, 219, 250
protein xxxv, xxxvi, 3, 5, 7, 8, 10, 15, 17, 18, 19, 20, 21, 23, 24, 25, 31, 33, 34, 36, 39, 40, 49, 52, 54, 57, 58, 65, 68, 81, 84, 85, 88, 89, 90, 91, 97, 101, 103, 104, 112, 113, 114, 115, 116, 117, 123, 124, 128, 133, 135, 140, 152, 157, 163, 164, 167, 173, 175, 186, 187, 202, 214, 216, 217, 222, 228, 231, 235, 238, 239, 243, 245, 253, 254, 263, 266, 272, 274, 275, 276, 305
protein powder 35, 216, 253
proteus 222
protozoa 51, 222
psoriasis xx, xxxvii, 27, 71, 145, 170, 173, 174, 175, 197, 198, 199, 201, 207, 215, 280
psyllium seed husk 55, 261
pulsar xix, 329, 330, 347, 349, 357

R

Rapture of the Church 402
red meat 2, 13, 56, 60, 63, 69, 71, 74, 80, 92, 95, 101, 102, 103, 112, 160, 165, 171, 177, 178, 179, 180, 201, 203, 237, 239, 248, 266, 274, 292, 305, 306
Redshift Light v, xix, xxv, xxix, xxx, 313, 314, 319, 320, 324, 327, 331, 334, 338, 339, 344, 345, 356, 357, 360, 367
refractive index 335
remission xi, xii, xx, xxii, xxxii, xxxiii, xxxvii, 56, 170, 171, 172, 184, 195, 234, 236, 263, 403
retinitis pigmentosa 15, 92
retinopathy 15
reverse osmosis water 35, 253, 254, 255, 268
rheumatoid arthritis xx, xxvi, xxxii, 7, 27, 71, 85, 139, 145, 159, 170, 171, 173, 175, 187, 191, 197, 215, 220, 242, 264, 272, 276
rice bran 55, 98, 125, 249, 261, 271
Rieske Instant Atherosclerosis Cycle (IAC) 13
Rieske Instant Osteoporosis Cycle (IOC) 19, 92
Rieske Minimum Black Hole Mass 348
Rieske Spacetime Drag Coefficient xix, 325, 326, 327, 328, 332, 336, 357
running 23, 25, 78, 286, 301, 376

S

safflower 71, 125, 159, 219, 220, 249, 271, 274, 275
salmon 34, 42, 64, 243, 246, 248, 263, 266
Salvation Prayer 405
Samson 137

Index

saturated fat xxxv, 11, 13, 17, 62, 68, 69, 70, 71, 72, 73, 74, 76, 77, 80, 86, 103, 104, 109, 110, 120, 145, 158, 160, 164, 165, 203, 237, 238, 239, 266, 292, 306
Schiaparelli, Giovanni Virginio 300
Scott, Captain Robert Falcon 56, 66
scurvy 42, 63, 64, 66, 67
seal meat xxvi, 65, 66
sesame 125, 249, 271
sheep 39, 247, 267, 272, 379
silo 151
Sjogren's syndrome 197
Smart, Elizabeth 287
soda 150, 255
soft drink 150
soybean 34, 71, 123, 125, 159, 215, 219, 220, 246, 249, 271, 273, 274, 275
Space Shuttle Challenger 290
Space Shuttle Endeavour 291
spacetime xix, 323, 325, 326, 328, 329, 332, 335, 338, 339, 340, 356, 357, 362, 367, 368, 369, 372
Special Relativity 315, 317, 321
speed of light xix, 308, 314, 315, 316, 317, 318, 322, 323, 324, 325, 326, 327, 328, 331, 332, 333, 334, 335, 336, 338, 339, 347, 356, 357, 362, 367, 369, 371, 372
sports xxxii, 23, 25
staphylococcus 224
starchy-vegetables 2, 6, 235
Starting Diet 171, 177, 234, 235, 245, 246, 250, 254, 265, 269, 276
Starting Diet Overview 171, 235
steers 7, 101, 240
Stefansson
   Vilhjalmur xiii, 42, 43, 65, 66, 67, 89, 99, 100, 266, 285
stool 45, 46, 48, 49, 50, 51, 55, 96, 98, 205, 206, 207, 223, 224, 232, 262
streptococcus 224

submucosa 52, 222
sucrose 6, 10, 135, 141, 143, 147, 150, 156, 270
sugar 4, 6, 8, 9, 10, 11, 15, 17, 19, 20, 21, 26, 28, 36, 38, 39, 64, 65, 76, 78, 79, 83, 84, 86, 93, 98, 112, 115, 119, 120, 128, 135, 141, 142, 143, 144, 146, 147, 148, 149, 150, 152, 156, 157, 162, 217, 245, 246, 248, 254, 255, 261, 266, 270, 271, 272, 273, 306
sugarcane 141
sunflower 71, 125, 159, 220, 249, 271
Swiss Alps 75

T

tea seed 125
telescope 323, 327, 331, 332, 340, 353, 363
Ten Commandments (Law) 383
testimonies 28, 44, 176, 177, 181, 202, 232, 302, 390, 392
Thailand 76
theanderose 6, 270
Theory of General Relativity 321
Three Mile Island Nuclear Power Station 291
tigers 58
time dilation 326
tomato seed 125, 249, 271
trans fats 72, 103, 147, 156, 158, 164, 167, 245, 271, 272, 273, 274
transit time 50, 51
triglycerides 11, 13, 17, 32, 72, 90, 120, 238, 249
turanose 6, 270
turkey 107, 246, 311
Type II diabetes 12, 60

Index

U

ulcerative colitis xi, xx, xxxii, xxxiii, xxxvii, 44, 50, 51, 85, 87, 95, 145, 170, 173, 174, 175, 184, 185, 187, 188, 215, 219, 236, 264, 276, 306, 308
universe iv, xvii, xix, xxix, xxx, xxxi, 282, 285, 301, 305, 315, 317, 318, 319, 320, 322, 324, 325, 326, 327, 328, 329, 331, 332, 334, 337, 338, 339, 340, 341, 342, 343, 344, 345, 347, 348, 349, 350, 351, 352, 355, 356, 357, 359, 363, 364, 365, 367, 369, 371
uric acid 94, 136
urine 31, 91, 113, 206
USDA Food Guide Pyramid x, xi, xxvi, 4, 7, 14, 15, 16, 22, 38, 60, 79, 83, 92, 95, 103, 112, 120, 164, 165, 171, 175, 194, 202, 203
USFDA Recommended Daily Allowance (RDA) 38, 60

V

vegetarians 30, 39, 55, 56, 62, 67, 72, 83, 85, 101, 103, 112, 113, 120, 152, 158, 160, 165, 167, 203, 204, 218, 239, 240, 295, 303
villi 52, 186, 222
viruses 51, 232, 234, 268
vitamin A 114, 199, 295
vitamin C 42, 63, 66, 67, 255
vitamin D 114, 117, 126, 268, 295

vitamins 6, 8, 35, 39, 55, 62, 63, 81, 98, 103, 111, 114, 142, 148, 163, 207, 218, 239, 243, 244, 245, 246, 250, 255, 256, 275, 305
vitamins, whole food 260
vulgaris 222, 280

W

Wallach, Dr. Joel D. DVM, ND. 133
walnut, 125, 206, 249, 255, 271
Washington, George, President 133, 163
whale 63
wheat germ 125, 249, 271
whey amino acid protein powder 35
White, Mrs. Ellen G. 152
wolf 43, 63, 293
wolves 58
Wormhole Theory 367, 368
worship
  animal 298, 299
  Earth 298, 299
Wright brothers xxii
Wright, Dr. Jonathan V., M.D. 216, 231

Y

yeast xx, 43, 51, 52, 204, 205, 206, 215, 223, 245, 247, 252, 255, 265, 267, 269, 271, 272, 274
yogurt 20, 75, 88, 121, 179, 237, 247, 248, 265, 267, 272
Young Earth Creationists 370